AF903189

Lecture Notes in Statistics

Vol. 1: R. A. Fisher: An Appreciation. Edited by S. E. Fienberg and D. V. Hinkley. xi, 208 pages, 1980.

Vol. 2: Mathematical Statistics and Probability Theory. Proceedings 1978. Edited by W. Klonecki, A. Kozek, and J. Rosiński. xxiv, 373 pages, 1980.

Vol. 3: B. D. Spencer, Benefit-Cost Analysis of Data Used to Allocate Funds. viii, 296 pages, 1980.

Vol. 4: E. A. van Doorn, Stochastic Monotonicity and Queueing Applications of Birth-Death Processes. vi, 118 pages, 1981.

Vol. 5: T. Rolski, Stationary Random Processes Associated with Point Processes. vi, 139 pages, 1981.

Vol. 6: S. S. Gupta and D.-Y. Huang, Multiple Statistical Decision Theory: Recent Developments. viii, 104 pages, 1981.

Vol. 7: M. Akahira and K. Takeuchi, Asymptotic Efficiency of Statistical Estimators. viii, 242 pages, 1981.

Vol. 8: The First Pannonian Symposium on Mathematical Statistics. Edited by P. Révész, L. Schmetterer, and V. M. Zolotarev. vi, 308 pages, 1981.

Vol. 9: B. Jørgensen, Statistical Properties of the Generalized Inverse Gaussian Distribution. vi, 188 pages, 1981.

Vol. 10: A. A. McIntosh, Fitting Linear Models: An Application of Conjugate Gradient Algorithms. vi, 200 pages, 1982.

Vol. 11: D. F. Nicholls and B. G. Quinn, Random Coefficient Autoregressive Models: An Introduction. v, 154 pages, 1982.

Vol. 12: M. Jacobson, Statistical Analysis of Counting Processes. vii, 226 pages, 1982.

Vol. 13: J. Pfanzagl (with the assistance of W. Wefelmeyer), Contributions to a General Asymptotic Statistical Theory. vii, 315 pages, 1982.

Vol. 14: GLIM 82: Proceedings of the International Conference on Generalised Linear Models. Edited by R. Gilchrist. v, 188 pages, 1982.

Vol. 15: K. R. W. Brewer and M. Hanif, Sampling with Unequal Probabilities. vii, 164 pages, 1983.

Lecture Notes in Statistics

Edited by D. Brillinger, S. Fienberg, J. Gani,
J. Hartigan, and K. Krickeberg

15

K.R.W. Brewer
Muhammad Hanif

Sampling With Unequal Probabilities

Springer Science+Business Media, LLC

K.R.W. Brewer
c/o Commonwealth Schools Commission
P.O. Box 34
Woden, Canberra
A.C.T. 2606
Australia

Muhammad Hanif
Department of Statistics
El-Fateh University
Tripoli
Libya (S.P.L.A.J.)

AMS Subject Classification: 62D05

Library of Congress Cataloging in Publication Data
Brewer, K. R. W.
Monograph on sampling with unequal probabilities.
(Lecture notes in statistics; v. 15)
Bibliography: p.
Includes indexes.
1. Sampling (Statistics) 2. Estimation theory.
I. Hanif, Muhammad. II. Title. III. Series: Lecture notes in statistics (Springer-Verlag); v. 15.
QA276.6.B74 1982 519.5'2 82-19256

With 9 Illustrations

© 1983 by Springer Science+Business Media New York
Originally published by Springer-Verlag New York Inc. in 1983.

All rights reserved. No part of this book may be translated or reproduced in any form without written permission from Springer Science+Business Media, LLC

9 8 7 6 5 4 3 2 1

ISBN 978-0-387-90807-6 ISBN 978-1-4684-9407-5 (eBook)
DOI 10.1007/978-1-4684-9407-5

PREFACE

Work for this monograph on sampling with unequal probabilities was started when Muhammad Hanif was a visitor to the then Commonwealth Bureau of Census and Statistics, Canberra, in 1969. It remained in abeyance until he again visited Canberra, this time the Australian National University's Survey Research Centre in 1978 as Visiting Fellow. The work was substantially completed when K.R.W. Brewer visited El-Fateh University during January 1980 as Visiting Professor. Finally, in 1982 the Bibliography was revised and corrected, and a number of references added which do not appear in the text. These are indicated by an asterisk (*).

The authors are indebted to Mr. E.K. Foreman and the sampling staff (past and present) at the Australian Bureau of Statistics for their help and encouragement and to Mrs Barbara Geary for her excellent mathematical typing.

Canberra

May 1982.

K.R.W. Brewer

Muhammad Hanif

CONTENTS

CHAPTER 1

AN INTRODUCTION TO SAMPLING WITH UNEQUAL PROBABILITIES

1.1 SOME BASIC CONCEPTS

Most survey work involves sampling from finite populations. There are two parts to any sampling strategy. First there is the selection procedure, the manner in which the sample units are to be selected from the finite population. Then there is the estimation procedure which prescribes how inferences are to be made from the sample to the population as a whole. These inferences may be either enumerative or analytical.

Enumerative inference seeks only to describe the particular finite population under study; analytical inference in some sense to explain it. Thus in dealing with a population of households we might attempt to enumerate the mean number of persons per household, the total number of persons in the population, the proportion of adults with tertiary education, the mean household income, and the ratio of total income to total number of persons (that is, per capita income). Enumerative inference typically concerns itself with such means, totals, proportions and ratios.

Viewing the same population analytically we might seek to regress household income on such variables as number of employed adults, educational level of household head, and dummy variables indicative of geographical location. Regression implies an explanatory model, and analytical inference consists of specifying what model is appropriate, estimating its parameters, and (at least in principle) checking to see whether the fitted model adequately describes the sample, and hence the population from which it was selected. Estimation of simple and multiple correlation coefficients also implies a regression model.

It is customary to distinguish between enumerative and analytical inference in terms of the complexity of the population characteristics being estimated. Estimating a mean is regarded as an enumerative problem; estimating a regression or correlation coefficient as an analytical one. But if the mean in question is a parameter of a simple explanatory model, the inference is in fact analytical. Similarly if the regression or correlation coefficient is being used as a purely descriptive measure without any explanatory significance, the inference is enumerative. (One may question the usefulness of estimating such coefficients in these circumstances, but not the fact that such estimation is often carried out.)

It is therefore the use of a statistical model for explanatory purposes which separates analytical from enumerative inference. (Such models may be used, especially at the sample design stage, to ensure that a particular piece of enumerative inference can be carried out with reasonable efficiency, but unless that enumerative inference is robust against model breakdown, it will not receive general acceptance.)

Analytical and enumerative inferences therefore proceed along entirely different paths. For analytical inference the model used provides its own probability structure, in terms of which the inference can proceed. This is the same methodology as is used in almost every other area of statistics. For enumerative inference a quite different probability structure is generally used, which depends on the manner in which the sample is selected. This is the classical finite population sampling inference deriving from Bowley (1913) and developed by Neyman (1934).

The co-existence of these two methods of inference is a matter that impinges very sharply on samples drawn with unequal probability. As long as all samples are drawn with equal probabilities the two forms of inference are equivalent (apart from the finite population correction) or can be made equivalent. Thus, for instance, in a stratified sampling situation where the sampling fractions differ from stratum to stratum, the two inferences can be made equivalent by considering each stratum as a separate population. A self-weighting sample drawn in two stages can also be considered, from the standpoint of analytical inference, as though it were an experimental design problem with the block effects being random variables.

But when each unit in a population has its own individual probability of inclusion in sample, there is no simple way in which the two methods of inference can be made to coincide. For analytical purposes the probabilities of selection are irrelevant. For enumerative inference they are crucial. This monograph is basically concerned with the situation where the probabilities of selection are relevant, and therefore with enumerative inference. Population models will be called on for comparative purposes for help in optimal design, and to shed light on the estimation process; but they will not be central to the problem, and any estimation of their parameters will be a means to an end rather than an end in itself.

The use of unequal probabilities in sampling was first suggested by Hansen and Hurwitz (1943). Prior to that date there had been substantial developments in sampling theory and practice, but all these had been based on the assumption that the probabilities of selection within each stratum would be equal. Hansen and Hurwitz demonstrated, however, that the use of unequal selection probabilities within a stratum frequently made for more efficient estimators of total than did equal probability sampling. They proposed a two-stage sampling scheme. The first-stage selection took place in independent draws. At each draw a single first-stage unit was selected with probabilities proportional to a size measure (the number of second-stage units within each first-stage unit). At the second stage, the same number of second-stage units was selected from each sample first-stage unit.

Because it was possible for the same first-stage unit to be selected more than once, this type of unequal probability sampling (*ups*) is generally known as *sampling with replacement*. Since, however, the independence of the draws is not a necessary condition for the units to have a non-zero probability of being selected more than once, another name first suggested by Hartley and Rao (1962) will be used instead. This is *multinomial sampling*, a term justified by the multinomial distribution of the number of units in the sample.

This scheme compared favourably with two other two-stage sampling schemes; these used unequal probabilities of selection at the first stage, and then took either a fixed number or a constant proportion of sub-sampling units from each selected first stage unit.

This first suggestion for the use of *ups* was thus already associated with the technique of multistage sampling and sampling with probability proportional to size (also known as *pps* or sometimes as *πps* sampling). Unequal probability sampling can, however, be used in single stage designs, and need not necessarily be with probability exactly proportional to size, although some kind of size measure is almost always used as a starting off point for assigning selection probabilities. A number of without replacement selection procedures result in probabilities of inclusion in sample which are only approximately proportional to size, and sometimes it is actually desirable to select with probability proportional to a power of size.

Practically all the published literature on *ups* to date has been concerned either with *sampling with replacement* (*multinomial sampling*) or *sampling without replacement*. It is these two cases which will be considered in the main portion of this Chapter. Chapters 2, 3 and 4 will be taken up with a detailed discussion of procedures for *sampling without replacement*. A full description of each procedure will be found in Chapter 2. Those procedures specifically designed for use with the Horvitz-Thompson (1952) estimator will be compared in Chapter 3, and other procedures in Chapter 4. In Chapter 5, *ups* is considered in the context of multistage sampling, and an algorithm is presented for estimating the components of variance for any sampling method that uses unbiased estimation. Finally, in Chapter 6, robust estimation is examined in the context of *ups*, and in particular the simultaneous optimization of the estimator and the selection probabilities in a large single stage sample.

1.2 NOTATION AND ABBREVIATIONS

The following is a summary of the principal notation to be used in this monograph. It is intended for reference purposes only, and further notation will also be introduced as required.

Capital letters and subscripts will be used for population values, and lower case for sampling values. Thus Y_I is the value of the Y-variate taken by the Ith population unit and y_i is the corresponding value taken by the ith sample unit. This departure from the usual practice is needed to avoid ambiguities in multistage sampling and in any case makes for greater clarity. The adoption of this notation is also useful for teaching purposes.

N, n, ... will denote the number of units in the population and sample respectively. Wherever the number of sample units is a random variable it will be denoted by n, and its expectation by ν.

Y, y, ... will denote values of the current or estimand variable.

X, x, ... will denote values of a benchmark variable used in ratio estimation.

Z, z, ... will denote measures of size.

P, p, ... will denote normed measures of size, that is, $P_I = Z_I/Z$ where

$$Z = \sum_{I=1}^{N} Z_I .$$

π_I ... will denote the probability of inclusion in sample of the Ith unit and π_i will denote its sample value.

π_{IJ} ... will denote the joint probability of inclusion in sample of the Ith and Jth population units and π_{ij} will denote its sample value.

E ... is the sampling expectation operator (expectation over all possible samples).

V ... is the sampling variance operator, and v will denote a sample estimator of the corresponding variance.

The following abbreviations will also be used:

srswr ... simple random sampling with replacement.

srswor ... simple random sampling without replacement.

πpswor ... probability of inclusion in sample proportional to size without replacement.

ppswr ... probability of selection proportional to size with replacement (multinomial sampling).

ups ... unequal probability sampling.

upswor ... unequal probability sampling without replacement.

upswr ... unequal probability sampling with replacement.

The last three abbreviations are more general in their application than the others in that they do not require the probabilities of selection to be proportional to a given size measure. They therefore embrace such schemes as sampling with probability proportional to a function of size, and the special selection procedures devised by Raj, by Das, and by Rao, Hartley and Cochran, mentioned in Section 1.5 below.

1.3 MULTINOMIAL SAMPLING (*Sampling with Replacement*)

As already mentioned, the sampling procedure of Hansen and Hurwitz (1943) used sampling *ppswr*. One unit was selected at each of n draws. The probability of selection of the Ith population unit at any of these draws was $P_I = Z_I/Z$, where Z_I was the measure of size of that unit and $Z = \sum_{I=1}^{N} Z_I$ was the total measure of size of the N units in the population.

The Hansen-Hurwitz estimator y'_{HH} of the population total, $Y = \sum_{I=1}^{N} Y_I$ is

$$y'_{\text{HH}} = \frac{1}{n} \sum_{i=1}^{n} \frac{y_i}{p_i} . \tag{1.3.1}$$

The estimator is unbiased and its variance may be expressed variously as

$$V(y'_{\text{HH}}) = \frac{1}{n} \left(\sum_{I=1}^{N} \frac{Y_I^2}{P_I} - Y^2 \right) , \tag{1.3.2}$$

$$= \frac{1}{n} \sum_{I=1}^{N} P_I \left(\frac{Y_I}{P_I} - Y \right)^2 , \text{ or} \tag{1.3.3}$$

$$= \frac{1}{2n} \sum_{\substack{I,J=1 \\ J \neq I}}^{N}\sum P_I P_J \left(\frac{Y_I}{P_I} - \frac{Y_J}{P_J} \right)^2 . \tag{1.3.4}$$

An unbiased estimator of this variance is

$$v\left(y'_{\text{HH}}\right) = \frac{1}{n(n-1)} \sum_{i=1}^{n} \left(\frac{y_i}{p_i} - y'_{\text{HH}}\right)^2 , \tag{1.3.5}$$

$$= \frac{1}{2n^2(n-1)} \sum_{\substack{i,j=1 \\ j \neq i}}^{n}\sum \left(\frac{y_i}{p_i} - \frac{y_j}{p_j}\right)^2 . \tag{1.3.6}$$

This scheme of *sampling with replacement* is less efficient than certain alternative schemes which have since been proposed for sampling without replacement. In spite of this some people prefer to use *sampling with replacement* for the following reasons:

(i) Selection of the sample is simple.

(ii) The method can be used for any finite predetermined (but not necessarily distinct) number of units in sample.

(iii) The unbiased estimator of variance is simple.

(iv) It is also comparatively easy to obtain unbiased estimators of total variance and components of variance in multistage designs.

1.4 πps*wor* METHODS USING THE HORVITZ-THOMPSON ESTIMATOR

Six years after Hansen and Hurwitz's paper, Madow (1949) proposed the use of systematic sampling with unequal probabilities so as to avoid the possibility of units being selected more than once. This suggestion was followed up by a large number of alternative selection procedures commencing with that devised by Narain (1951), which will be considered in detail in Chapter 2.

Horvitz and Thompson (1952) produced a general theory of sampling with unequal probabilities without replacement based on the use of the estimator

$$y'_{\text{HT}} = \sum_{i=1}^{n} \frac{y_i}{\pi_i} , \tag{1.4.1}$$

where y_i is the value of the ith sample unit. The important properties of the Horvitz-Thompson estimator are as follows:

(i) It is the only unbiased estimator of the class in which the same weight is attached to a particular population unit whenever it is selected (Horvitz and Thompson 1952).

(ii) It is admissible in the class of all homogeneous linear unbiased estimators of population total Y ; that is, there does not exist any member of that class which has a smaller variance than y'_{HT} (Roy and Chakravarti, 1960; Godambe, 1960).

(iii) If the Y_I are all exactly proportional to the corresponding π_I and the number of units in sample is fixed, the variance of y'_{HT} is zero. This is a property usually associated with ratio estimators and will be referred to as the *ratio estimator property*.

If Z_I values are known for all units in the population, and the Y_I are approximately proportional to the Z_I, the variance of y'_{HT} can be made small by setting the π_I proportional to the Z_I. This is one important reason why selection with probability proportional to size has assumed a central importance in *ups*.

The following is a summary of the general estimation theory for selection with probabilities to size without replacement (*πpswor*) based on the papers by Horvitz and Thompson (1952) and by Hanurav (1962a, 1967).

$$\sum_{I=1}^{N} \pi_I = E(n) = \nu \ , \tag{1.4.2}$$

$$\sum_{J \neq I}^{N} \pi_{IJ} = (\nu-1)\pi_I \ , \tag{1.4.3}$$

and

$$\sum_{\substack{I,J=1 \\ J \neq I}}^{N}\sum \pi_{IJ} = \nu(\nu-1) + V(n) \ . \tag{1.4.4}$$

The variance of the unbiased estimator y'_{HT} is:

$$V\left(y'_{\mathrm{HT}}\right) = \sum_{I=1}^{N} \frac{1-\pi_I}{\pi_I} Y_I^2 + \sum_{\substack{I,J=1 \\ J \neq I}}^{N}\sum \frac{\pi_{IJ}-\pi_I\pi_J}{\pi_I\pi_J} Y_I Y_J \ . \tag{1.4.5}$$

The following alternative expression, which was derived independently by Sen (1953) and by Yates and Grundy (1953) (it will be attributed to Sen-Yates-Grundy), is valid only if the number of units in sample is fixed:

$$V\left(y'_{\mathrm{HT}}\right) = \sum_{\substack{I,J=1 \\ J > I}}^{N}\sum \left(\pi_I\pi_J-\pi_{IJ}\right)\left(\frac{Y_I}{\pi_I} - \frac{Y_J}{\pi_J}\right)^2 \ . \tag{1.4.6}$$

The following estimator of $V\left(y'_{\mathrm{HT}}\right)$ is unbiased if $\pi_{IJ} \neq 0$ for all $J \neq I$, that is if all possible pairs of distinct population units have a non-zero probability of inclusion in sample. (Such estimators as these will be referred to as *conditionally unbiased*.)

$$v(y'_{HT}) = \sum_{i=1}^{n} \frac{1-\pi_i}{\pi_i^2} y_i^2 + \sum_{\substack{i,j=1 \\ j \neq i}}^{n}\sum \frac{\pi_{ij}-\pi_i\pi_j}{\pi_{ij}\pi_i\pi_j} y_i y_j . \quad (1.4.7)$$

This estimator suffers from the disadvantage that it is not always zero when the variance is zero. The following alternative conditionally unbiased estimator was suggested by Sen (1953) and by Yates and Grundy (1953) for use when the number of sample units is fixed:

$$v_{SYG}(y'_{HT}) = \sum_{\substack{i,j=1 \\ j>i}}^{n}\sum \frac{\pi_i\pi_j-\pi_{ij}}{\pi_{ij}} \left(\frac{y_i}{\pi_i} - \frac{y_i}{\pi_j}\right)^2 . \quad (1.4.8)$$

Both (1.4.7) and (1.4.8) can assume negative values, but (1.4.8) rarely seems to do so in practice. It has also performed much better than (1.4.7) in a number of empirical comparisons, commencing with that in Yates and Grundy's (1953) paper. For $n = 2$ it is the only possible non-negative variance estimator (Vijayan, 1975).

Sen (1953) also compared the efficiency of (1.4.7) and (1.4.8) taking a population of five units and selecting all possible samples of two units under the following two schemes:

(a) The first unit is drawn with *pps* and the second unit with *pps* without replacement.

(b) The first unit is drawn with *pps* and the second unit with equal probabilities without replacement.

He demonstrated that the expression (1.4.8) took positive values for all these samples, but that (1.4.7) was negative for some of the pairs. He further showed that $\pi_{ij} > \pi_i\pi_j$; for all $i \neq j$ for $n = 2$, and hence that when selection is made with *πpswor* using the Horvitz-Thompson estimator, (1.4.8) is always positive.

Raj (1956a) proved further that the expression (1.4.8) was always positive under the schemes (a) and (b) above.

Rao (1963a) also proved under two other well-known procedures for *upswor* that (1.4.8) was always positive. These procedures (that of Midzuno - reported by Horvitz and Thompson (1952) - and the Yates and Grundy (1953) rejective procedure) will be described in Chapter 2.

Rao and Singh (1973) used Brewer's (1963) *πpswor* selection procedure to compare (1.4.7) and (1.4.8) for the case $n = 2$, employing a wide variety of populations. Their empirical evidence also indicates that (1.4.8) is more stable than (1.4.7). A similar result was obtained by Lanke (1974) using Hajek's "Method I" (1964b).

1.5 *upswor* METHODS USING OTHER ESTIMATORS

Das (1951) suggested the following strategy. A sample of n units is selected. At each draw the selection is made among those units not already selected with probabilities proportional to size. The estimator t'_{mean} of the population total Y is the arithmetic mean of the n unbiased estimators:

$$\left.\begin{aligned} t'_1 &= \frac{y_1}{p_1}\,, \\ t'_2 &= \frac{1-p_1}{p_1 p_2}\frac{y_2}{N-1}\,, \\ &\cdots \\ t'_r &= \prod_{i=1}^{r-1}\left(1-\sum_{j=1}^{i} p_j\right) y_r \div \prod_{i=1}^{r} p_i \prod_{i=1}^{r-1}(N-i)\,, \quad r = 1, 2, \ldots, n\,. \end{aligned}\right\} \tag{1.5.1}$$

Raj (1956a) modified Das's strategy as follows. The estimator t_{mean} of the population total Y is the arithmetic mean of the following set of unbiased and uncorrelated estimators:

$$\left.\begin{aligned} t_1 &= \frac{y_1}{p_1}\,, \\ t_2 &= y_1 + \frac{y_2}{p_2}\left(1-p_1\right)\,, \\ &\cdots \\ t_k &= y_1 + y_2 + \cdots + y_{n-1} + \frac{y_n}{p_n}\left(1 - p_1 - p_2 - \cdots - p_{n-1}\right)\,. \end{aligned}\right\} \tag{1.5.2}$$

(The estimator t_1 thus depends only on the first unit selected, t_2 on the first two units, and so on.)

Murthy (1957) considered all possible permutations of the sample order which led to different values of t_{mean}. He proved that the symmetrized estimator, which gave weight to each possible value of t_{mean} in proportion to the *a priori* probability of the observed sample units being selected in that order, had a smaller variance than t_{mean}.

Rao, Hartley and Cochran's (1962) sampling strategy is as follows. The population units are divided randomly into n groups of N_J units, $J = 1, 2, 3, \ldots, n$, where the N_J are predetermined. One unit is selected from each group, the probabilities of selection being the normed measures of size within the group. Their estimation procedure is to form the Horvitz-Thompson estimator for each group separately and to add these over the groups.

1.6 LIST OF PROCEDURES FOR SAMPLING WITHOUT REPLACEMENT

The following is a list of 50 *πpswor* procedures. A mnemonic title is suggested for each, and a basic reference is given. Descriptions of these procedures will be given in Chapter 2.

Procedure 1: Ordered Systematic Procedure; Madow (1949).
Procedure 2: Random Systematic Procedure; Goodman and Kish (1950).
Procedure 3: Grundy's Systematic Procedure; Grundy (1954).
Procedure 4: Yates-Grundy Draw-by-Draw Procedure; Yates and Grundy (1953).
Procedure 5: Yates-Grundy Rejective Procedure; Yates and Grundy (1953).
Procedure 6: Midzuno's Procedure; reported by Horvitz and Thompson (1952).
Procedure 7: Narain's Procedure; Narain (1951).
Procedure 8: Brewer's Procedure; Brewer (1963, 1975).
Procedure 9: Durbin's "Method I"; Durbin (1967).
Procedure 10: Durbin's "Grouped Method"; Durbin (1967).
Procedure 11: Rao-Sampford Procedure; Rao (1965), Sampford (1967).
Procedure 12: Durbin-Sampford Procedure; Sampford (1967).
Procedure 13: Fellegi's Procedure; Fellegi (1963).
Procedure 14: Carroll-Hartley Rejective Procedure; Carroll and Hartley (1964).
Procedure 15: Carroll-Hartley Draw-by-Draw Procedure; Carroll and Hartley (1964).
Procedure 16: Carroll-Hartley Whole Sample Procedure; Carroll and Hartley (1964).
Procedure 17: Durbin-Hanurav Procedure; Durbin (1953b); Hanurav (1966, 1967).
Procedure 18: Hanurav's Scheme B-A'; Hanurav (1967).
Procedure 19: Hanurav-Vijayan Procedure; Hanurav (1967); Vijayan (1968).
Procedure 20: Raj's Variance Minimization Procedure; Raj (1956b).
Procedure 21: Hanurav's Simple Junctional Procedure; Hanurav (1962a).
Procedure 22: Hanurav's Modified Junctional Procedure; Hanurav (1962a).
Procedure 23: Hanurav's Double Junctional Procedure; Hanurav (1962a).

Procedure 24: Hanurav's Sequential Procedure; Hanurav (1962a).

Procedure 25: Rao-Hartley-Cochran Procedure; Rao, Hartley and Cochran (1962).

Procedure 26: Stevens' Procedure; Stevens (1958).

Procedure 27: Poisson Sampling; Hajek (1964b).

Procedure 28: Hajek's "Method I"; Hajek (1964b).

Procedure 29: Hajek's "Method II"; Hajek (1964b).

Procedure 30: Hajek's "Method III"; Hajek (1964b).

Procedure 31: Hajek's "Method IV"; Hajek (1964b).

Procedure 32: Deming's Systematic Procedure; Deming (1960).

Procedure 33: Variance Estimator Optimization Procedure; Brewer and Hanif (1969a).

Procedure 34: Jessen's "Method 1"; Jessen (1969).

Procedure 35: Jessen's "Method 2"; Jessen (1969).

Procedure 36: Jessen's "Method 3"; Jessen (1969).

Procedure 37: Jessen's "Method 4"; Jessen (1969).

Procedure 38: Modified Poisson Sampling; Ogus and Clark (1971).

Procedure 39: Collocated Sampling; Brewer, Early and Hanif (1980).

Procedure 40: Das-Mohanty Procedure; Das and Mohanty (1973).

Procedure 41: Mukhopadhyay's Procedure; Mukhopadhyay (1972).

Procedure 42: Sinha's Extension Procedure; Sinha (1973).

Procedure 43: Sinha's Reduction Procedure; Sinha (1973).

Procedure 44: Chaudhuri's Procedure; Chaudhuri (1976).

Procedure 45: Lahiri's Procedure; Lahiri (1951).

Procedure 46: Ikeda-Midzuno Procedure; Midzuno (1952).

Procedure 47: Fuller's "Scheme B"; Fuller (1971).

Procedure 48: Singh's Procedure; Singh (1978).

Procedure 49: Choudhry's Procedure; Choudhry (1979).

Procedure 50: Chromy's Procedure; Chromy (1979).

1.7 CLASSIFICATION OF PROCEDURES

The fifty procedures listed above may be classified in a number of ways. This section deals with three of the most useful and instructive classifications.

1.7.1 Classification by Manner of Selection

The classification by manner of selection set out below is based on that of Carroll and Hartley (1964), which although not entirely unambiguous is nevertheless useful for expository purposes.

(i) Draw-by-Draw Procedures

At each successive draw one unit is selected, usually from among those population units not previously selected. Probabilities of selection are defined for each draw and (since the selection is without replacement) always depend on which units are already selected. If the probabilities of selection at a given draw are (apart from a normalizing factor) independent of which units were selected at previous draws, they are sometimes referred to as *working probabilities*. The draw-by-draw procedures listed above are Procedures 4, 6, 7, 8, 9, 10, 12, 13, 15, 18, 19, 21, 22, 23, 24, 25, 26, 41, 44, 46, 47, 49 and 50.

(ii) Systematic Procedures

Systematic selection involves an ordering of the population and the cumulation of inclusion probabilities. The order of units may or may not be random. A random number r $(0 < r \leq 1)$ is chosen and the n units selected are those whose cumulated values of π_I (the desired probability of inclusion) are the smallest equal to or greater than each of $r, r+1, r+2, \ldots, r+n-1$. The systematic procedures listed above are Procedures 1, 2, 3, 32 and 48.

(iii) Rejective Procedures

The term *rejective* has been employed by Hajek (1964b) and is somewhat wider in its connotation than the term *mass draw* used by Carroll and Hartley (1964). Rejective procedures resemble draw-by-draw procedures in that only a single unit is selected at each of n successive draws. They differ from ordinary draw-by-draw procedures in that the selection at a given draw may give rise to the selection of an already selected unit, in which case the partial sample is abandoned and the selection recommenced. The rejective procedures listed above are Procedures 5, 11, 14, 17, 28, 29, 30 and 31.

(iv) Whole Sample Procedures

In these procedures the units are not individually drawn: a probability is specified for each possible sample of n distinct units and one selection using these probabilities selects the whole sample. The whole sample procedures listed above are Procedures 16, 20, 33, 34, 35, 36, 37, 40 and 45.

(v) Other Selection Procedures

Other selection procedures not listed in the above four categories are as follows:

Procedure 27: Poisson Sampling.

Procedure 38: Modified Poisson Sampling.

Procedure 39: Collocated Sampling.

Procedure 42: Sinha's Extension Procedure.

Procedure 43: Sinha's Reduction Procedure.

1.7.2 Classification by Equivalence Class

Two procedures belong to the same equivalence class when the joint probabilities of inclusion of all possible combinations of units are identical. It is obvious that each systematic, draw-by-draw and rejective procedure has an equivalent whole sample procedure. Godambe (1955) pointed out that any whole sample procedure also has a draw-by-draw equivalent. Hence it is possible, for a number of the procedures described in this monograph, to devise different selection procedures in the same equivalence class in a straightforward fashion.

Procedures 9, 11 and 12 belong to an equivalence class possessing the characteristic that the joint inclusion probabilities π_{ij} can be stated explicitly in analytic form, thus making the variance formulae comparatively simple. This will be referred to as Equivalence Class A. Procedure 8 belongs to this class for $n = 2$.

Procedures 14, 15 and 16 belong to a second equivalence class. Since they include the *symmetric mass draw* procedures, they may be designated as *symmetric* procedures, or Equivalence Class B. For the case $n = 2$, Procedures 13 and 49 also belong to this equivalence class.

1.7.3 Classification by Type of Estimator Appropriate

The first estimator suggested for use with sampling with unequal probabilities without replacement was that of Horvitz and Thompson (1.4.1). Any selection procedure may be used with the Horvitz-Thompson estimator, but it is only those for which the sample number is fixed and the probabilities of inclusion in sample can be made exactly proportional to an already known measure of size for which this estimator has the *ratio estimator property* (see Section 1.4). A further important property which applies under the same conditions is that the value of the Sen-Yates-Grundy (1953) variance estimator (1.4.8) is then also zero. Procedures possessing this property exactly are all those listed except Procedures 4, 5, 25-31, 38-39 and 45-46. Most of these exceptions were initially put forward as approximations to an ideal procedure which would confer the ratio estimator property on the Horvitz-Thompson estimator. Procedure 25 however was suggested explicitly for use with a special estimator. Certain other special estimators, notably t'_{mean}, t_{mean} and t_{symm} have since been suggested for use with Procedure 4 by Das (1951), Raj (1956a) and Murthy (1957). The special estimators t_{mean} and t_{symm} , when used in conjunction with Procedure 4 also possess the ratio estimator property.

All published selection procedures may therefore be classified into three groups:

(i) Those for which the Horvitz and Thompson estimator possesses the ratio estimator property exactly. These are Procedures 1-3, 6-24, 32-37, 40-44 and 47-50.

(ii) Those for which no estimator possesses this property exactly but for which the Horvitz and Thompson estimator possesses it approximately. These are Procedures

5, 26 and 28-31.

(iii) Those for which some other estimator or estimators possess this property.

Procedure 4: Yates-Grundy Draw-by-Draw Procedure - with the estimator t_{mean} suggested by Raj (1956a) or the estimator t_{symm} suggested by Murthy (1957).

Procedure 25: Rao-Hartley-Cochran Procedure - with RHC estimator.

Procedure 27: Poisson Sampling - with a ratio estimator (Brewer, Early and Joyce 1972).

Procedure 38: Modified Poisson Sampling - with a ratio estimator (Brewer, Early and Hanif 1980).

Procedure 39: Collocated Sampling - with a ratio estimator (Brewer, Early and Hanif 1980).

Procedure 45: Lahiri's Procedure - with the conventional ratio estimator.

Procedure 46: Ikeda-Midzuno Procedure - with the conventional ratio estimator.

1.8 SOME ASYMPTOTIC VARIANCE FORMULAE FOR *πpswor*

The variance (1.4.5) of the Horvitz-Thompson estimator involves the quantities π_{IJ}. Considerable difficulties are involved in the determination of these quantities for most of the procedures listed in Section 1.6. Hartley and Rao (1962) obtained an approximate solution to this problem for the Random Systematic Procedure with the help of an asymptotic theory which assumed that N was large and that $n/N \to 0$ as $N \to \infty$. They obtained an expression for the asymptotic value of π_{IJ} under these assumptions and substituted this in expression (1.4.5) to obtain the following asymptotic variance formula for any n :

$$V(y'_{\mathrm{HT}}) \doteq \sum_{I=1}^{N} \pi_I\left(1 - \frac{n-1}{n}\pi_I\right)\left(\frac{Y_I}{\pi_I} - \frac{Y}{n}\right)^2 - \frac{n-1}{n^2}\sum_{I=1}^{N}\left(2\pi_I^3 - \frac{\pi_I^2}{2}\sum_{J=1}^{N}\pi_J^2\right)\left(\frac{Y_I}{\pi_I} - \frac{Y}{n}\right)^2 + \frac{2(n-1)}{n^3}\left(\sum_{I=1}^{N}\pi_I Y_I - \frac{Y}{n}\sum_{J=1}^{N}\pi_J^2\right)^2 . \quad (1.8.1)$$

This formula is correct to order N^0 .

Rao (1963a) showed that with $n = 2$ the variances of the Horvitz-Thompson estimator for three selection procedures were given asymptotically to order N^0 by

$$V(y'_{\mathrm{HT}}) \doteq \sum_{I=1}^{N} \pi_I\left(1 - \frac{\pi_I}{2}\right)\left(\frac{Y_I}{\pi_I} - \frac{Y}{2}\right)^2 - \tfrac{1}{4}\sum_{I=1}^{N}\left(\pi_I^3 - \frac{\pi_I^2}{4}\sum_{J=1}^{N}\pi_J^2\right)\left(\frac{Y_I}{\pi_I} - \frac{Y}{2}\right)^2$$
$$+ \lambda\left(\sum_{I=1}^{N}\pi_I Y_I - \frac{Y}{2}\sum_{J=1}^{N} Y_J^2\right)^2 . \quad (1.8.2)$$

The value of λ in this formula is 3/32 for Narain's Procedure, 1/8 for the Carroll-Hartley Rejective Procedure and 1/4 for the Random Systematic Procedure. Rao (1965) later showed that $\lambda = 0$ for Brewer's Procedure. Since Fellegi's Procedure and the three Carroll-Hartley Procedures are in the same equivalence class, the value of λ is 1/8 for all these Procedures. Similarly, since the Rao-Sampford Procedure, Durbin's "Method I" and Brewer's Procedure are in the same equivalence class, $\lambda = 0$ for all these procedures.

All the procedures mentioned so far in this Section have the same variance formula to order N^1 for $n = 2$, namely:

$$V(y'_{\mathrm{HT}}) \doteq \sum_{I=1}^{N} \pi_I\left(1 - \frac{\pi_I}{2}\right)\left(\frac{Y_I}{\pi_I} - \frac{Y}{2}\right)^2 . \quad (1.8.3)$$

From (1.8.1) it follows that for the Random Systematic Procedure, to order N^1,

$$V(y'_{\mathrm{HT}}) \doteq \sum_{I=1}^{N} \pi_I\left(1 - \frac{n-1}{n}\pi_I\right)\left(\frac{Y_I}{\pi_I} - \frac{Y}{n}\right)^2 . \quad (1.8.4)$$

Equation (1.8.4) was also shown by Rao (1963b) to be asymptotically valid for Narain's Procedure and the Carroll-Hartley Rejective Procedure.

There are almost certainly other *πpswor* procedures for which (1.8.4) is also asymptotically valid, but it seems necessary to derive it for each one separately. It is shown in this Section that under the assumption of a linear stochastic model, (1.3.4) is asymptotically valid for all *πpswor* procedures (Hanif, 1974).

The model is the same as that employed by Cochran (1953), which appears to have been originated by Smith (1938). The same linear model has been employed by several other authors, for example, Rao (1966), T.J. Rao (1967), Hanurav (1967), Vijayan (1967), Foreman and Brewer (1971), Hanif and Ahmad (1977), Cassel, Särndal and Wretman (1977).

The assumption is that the observed population can be treated as a sample of one from an infinite set of hypothetical populations generated by a stochastic model. In this particular application of the model it is assumed that the probabilities of inclusion in sample π_I are exactly proportional to size and constant from population to population. The model specification then is

$$\left.\begin{aligned} &Y_I = \beta Z_I + \epsilon_I \, , \\ &E^*(\epsilon_I) = 0 \, , \quad E^*(\epsilon_I \epsilon_J) = \begin{cases} \sigma_I^2 \, , & J = I \\ 0 \, , & \text{otherwise} \end{cases} \\ &\sigma_I^2 = \sigma^2 Z_I^{2\gamma} \quad \text{where } \tfrac{1}{2} \le \gamma \le 1 \, , \end{aligned}\right\} \tag{1.8.5}$$

where β, σ^2 are constants and E^* denotes the expectation over all possible hypothetical populations. The model based variance of y' regarded as an estimator of Y will be written as

$$V^*(y'_{\text{HT}}) = E^*(y'_{\text{HT}} - Y)^2 \, . \tag{1.8.6}$$

The design expectation of expression (1.8.6) when $\pi_I \propto Z_I$ is

$$EE^*(y'_{\text{HT}} - Y)^2 = \sum_{I=1}^{N} \sigma_I^2 \left(\frac{1}{\pi_I} - 1\right) ; \tag{1.8.7}$$

but the same expression may be obtained asymptotically from the model expectation of the right side of the expression (1.8.4). This model expectation is

$$E^*\left[\sum_{I=1}^{N} \pi_I \left(1 - \frac{n-1}{n}\pi_I\right)\left(\frac{Y_I}{\pi_I} - \frac{Y}{n}\right)^2\right]$$

$$= \sum_{I=1}^{N} \sigma_I^2 \left(\frac{1}{\pi_I} - 1\right) + \frac{n-1}{n^2}\left\{2 \sum_{I=1}^{N} \sigma_I^2 \pi_I - \sum_{I=1}^{N} \frac{\pi_I^2}{n} - \sum_{I=1}^{N} \sigma_I^2\right\} . \tag{1.8.8}$$

Now the second term of the expression (1.8.8) is of order N^0 while the leading term contains only expressions of order N^2 and N^1. Hence asymptotically only the leading term is left, which is the same as (1.8.7). Hence under model (1.8.5) the asymptotic variance formula (1.8.4) is valid for all *πpswor* procedures.

1.9 A GENERAL THEORY OF ESTIMATORS POSSESSING THE RATIO ESTIMATOR PROPERTY

Vijayan (1975), Rao and Vijayan (1977), and Rao (1979) have presented some general results which enable the mean square error of any linear estimator of total possessing the ratio estimator property in a straightforward fashion, and also exhibit the necessary form of any non-negative quadratic unbiased estimators of that mean square error. These results are essentially contained in the following theorem from Rao (1979).

Let the general linear estimator of the population total Y, based on a sample

s with associated probability of selection $p(s)$ be written

$$\hat{Y} = \sum_{I=1}^{N} d_{Is} Y_I \; , \tag{1.9.1}$$

where the weights d_{Is} do not depend on the Y_I, and $d_{Is} = 0$ if $I \notin s$. Suppose $\hat{Y}$ is such that $\text{MSE}(\hat{Y})$ becomes zero when $Y_I = cZ_I$ where $c \neq 0$ is an arbitrary constant (the ratio estimator property). Then

(a) $\text{MSE}(\hat{Y})$ reduces to

$$\text{MSE}(\hat{Y}) = -\sum\sum_{\substack{I,J=1\\J>I}}^{N} d_{IJ} Z_I Z_J (R_I - R_J)^2 \; , \tag{1.9.2}$$

where $R_I = Y_I/Z_I$ and

$$d_{IJ} = E(d_{Is}-1)(d_{Js}-1) \; ; \tag{1.9.3}$$

(b) a non-negative quadratic unbiased estimator of $\text{MSE}(\hat{Y})$ is necessarily of the form

$$\text{mse}(\hat{Y}) = -\sum\sum_{\substack{I,J=1\\J>I}} d_{IJ}(s) Z_I Z_J (R_I - R_J)^2 \; , \tag{1.9.4}$$

where the coefficients $d_{IJ}(s)$ do not depend on the Y_I, $d_{IJ}(s) = 0$ if s does not contain both units I and J and the following unbiasedness condition is satisfied:

$$\sum_{s \ni I,J} p(s) d_{IJ}(s) = d_{IJ} \; , \quad J > I \; . \tag{1.9.5}$$

The proof of this theorem depends on the matrix lemma that if $B = (b_{IJ})$ is a positive semidefinite matrix and $\sum\sum_{I,J=1}^{N} b_{IJ} = 0$, then $\sum_{J=1}^{N} b_{IJ} = 0$ for all I.

(a) Using (1.9.3) we can write

$$\begin{aligned} \text{MSE}(\hat{Y}) = E(\hat{Y}-Y)^2 &= \sum\sum_{I,J=1}^{N} d_{IJ} Z_I Z_J R_I R_J \\ &\equiv \sum_{I=1}^{N} d_{II}^2 Z_I^2 R_I^2 + \sum\sum_{\substack{I,J=1\\J\neq I}}^{N} d_{IJ} Z_I Z_J R_I R_J \; . \end{aligned} \tag{1.9.6}$$

Since $\text{MSE}(\hat{Y}) = 0$ when all $R_I = c$, we get $\sum\sum_{I,J=1}^{N} d_{IJ} Z_I Z_J = 0$ and by the lemma $\sum_{J=1}^{N} d_{IJ}Z_IZ_J = 0$ for each I, that is, $d_{II}Z_I^2 = -\sum_{J\neq I}^{N} d_{IJ}Z_IZ_J$. Substituting this result in (1.9.6),

$$\begin{aligned}\text{MSE}(\hat{Y}) &= -\sum\sum_{\substack{I,J=1\\J\neq I}}^{N} d_{IJ}Z_IZ_J\left(R_I^2-R_IR_J\right)\\ &\equiv -\sum\sum_{\substack{I,J=1\\J>I}}^{N} d_{IJ}Z_IZ_J\left(R_I-R_J\right)^2 . \end{aligned} \tag{1.9.2}$$

(b) Any unbiased non-negative quadratic estimator may be written

$$\text{mse}(\hat{Y}) = \sum\sum_{I,J=1}^{N} d_{IJ}(s)Y_IY_J \tag{1.9.7}$$

$$\equiv -\sum\sum_{\substack{I,J=1\\J>I}}^{N} d_{IJ}(s)Z_IZ_J\left(R_I-R_J\right)^2 + \sum\sum_{I,J=1}^{N} d_{IJ}(s)Z_IZ_JR_I^2 . \tag{1.9.8}$$

By unbiasedness, non-negativity, and the ratio estimator property, $\sum\sum_{I,J=1}^{N} d_{IJ}(S)Z_IZ_J$ must be zero when all the $R_I = c$. But the $d_{IJ}(S)$ are independent of the Y_I, hence it must always be zero. Hence, by the lemma $\sum_{J=1}^{N} d_{IJ}(S)Z_IZ_J = 0$ for all I, and the second term on the right hand side of (1.9.8) disappears, leaving expression (1.9.4). //

Infinitely many choices for $d_{IJ}(s)$ satisfying (1.9.5) are possible. Rao (1979) makes two suggestions. The first is

$$d_{IJ}(s) = d_{IJ}/\pi_{IJ} \tag{1.9.9}$$

and the second, valid for unbiased $\hat{Y}$ only, is

$$d_{IJ}(s) = d_{Is}d_{Js} - f_{IJ}(s)/p(s) \tag{1.9.10}$$

where $f_{IJ}(s)$ is any choice satisfying $\sum_{s\ni I,J} f_{IJ}(s) = 1$, $J > I$. The second has the advantage that the d_{IJ} need not be calculated.

For the important special case where the sample is constrained to consist of two distinct units (I and J) the only possible non-negative unbiased estimator of $\text{MSE}(\hat{Y})$ is

$$\text{mse}(\hat{Y}) = -\frac{d_{IJ}}{\pi_{IJ}} Z_IZ_J\left(R_I-R_J\right)^2 . \tag{1.9.11}$$

For the Hansen-Hurwitz estimator used with *ppswr* the choice

$$d_{IJ}(s) = -\frac{1}{n}\frac{t_I t_J}{n(n-1)P_I P_J}, \tag{1.9.12}$$

where t_I is the number of times the Ith unit appears in sample, and $P_I = Z_I/Z$, leads to the usual variance estimator,

$$v\left(y'_{\text{HH}}\right) = \frac{1}{n(n-1)} \sum_{I\in s}{}' t_I \left(\frac{Y_I}{P_I} - y'_{\text{HH}}\right)^2 . \tag{1.9.13}$$

Alternatively (1.9.9) leads to Rao's new non-negative variance estimator

$$v_1\left(y'_{\text{HH}}\right) = \frac{1}{n} \sum\sum_{\substack{I,J\in s\\ J>I}} \frac{P_I P_J}{\pi_{IJ}} \left(\frac{Y_I}{P_I} - \frac{Y_J}{P_J}\right)^2 . \tag{1.9.14}$$

For the Horvitz-Thompson estimator used with fixed sample size *πpswor* (1.9.9) leads to the Sen-Yates-Grundy variance estimator (1.4.8) which for $n = 2$ is the only possible non-negative unbiased variance estimator. But (1.9.10) with $f_{IJ} = M_2^{-1}$ where $M_2 = \binom{N-2}{n-2}$, gives a new variance estimator for $n > 2$, that is

$$v_2\left(y'_{\text{HT}}\right) = \sum\sum_{\substack{I,J\in s\\ J>I}} \frac{\pi_I\pi_J - p(s)M_2}{p(s)M_2} \left(\frac{Y_I}{\pi_I} - \frac{Y_J}{\pi_J}\right)^2 . \tag{1.9.15}$$

Further, (1.9.10) with $f_{IJ}(s) = p(s|I, J)$ (the conditional probability of getting s given that I and J were selected in the first two draws) gives another new estimator for $n > 2$, that is

$$v_3\left(y'_{\text{HT}}\right) = \sum\sum_{\substack{I,J\in s\\ J>I}} \frac{\pi_I\pi_J p(s|I,J) - p(s)}{p(s)} \left(\frac{Y_I}{\pi_I} - \frac{Y_J}{\pi_J}\right)^2 . \tag{1.9.16}$$

The properties of (1.9.15) and (1.9.16) remain to be investigated, but (1.9.15) in particular appears to offer some gain in computational simplicity.

Rao (1969) also applied the above theory to Murthy's estimator and the Rao-Hartley-Cochran estimator, but without producing any new variance estimator of practical interest. The work of Rao and Vijayan (1977), in producing new variance estimators for the unbiased ratio estimator used with selection probabilities proportional to aggregative size, is considered in Chapter 4.

CHAPTER 2

DESCRIPTIONS OF PROCEDURES FOR SAMPLING WITH UNEQUAL PROBABILITIES WITHOUT REPLACEMENT

2.1 INTRODUCTION

In Chapter 1, 50 *upswor* procedures were listed. In this Chapter these selection procedures will be described in detail. The descriptions will be aimed at indicating the relationships between selection procedures and at showing how the disadvantages of some methods have led to the suggestion of others. The following descriptive abbreviations will be used.

strπps: probability of inclusion strictly proportional to size,

strwor: strictly without replacement,

n fixed: number of units in sample fixed,

syst: systematic,

d by d: draw by draw,

ws: whole sample,

ord: ordered,

unord: unordered.

In addition to these purely descriptive abbreviations certain disadvantages will be indicated as follows:

inexact: fails to satisfy at least one of the three descriptions *strπps*, *strwor* and *n fixed* above,

$n = 2$ only: limited to two sample units per stratum,

b est var: estimator of variance generally biased,

j p enum: calculation of joint probabilities of inclusion in sample involves enumeration of all possible selections, or at least a large number of them, *j p iter*: calculation of joint probabilities of inclusion in sample involves iteration on computer,

not gen app: not generally applicable,

non-rotg: non-rotating.

The last two of these disadvantages require some explanation.

(i) Not Generally Applicable

Since the probability of inclusion is proportional to size and no probability of inclusion can be greater than unity, the theoretical limit to the size of individual units is Z/n. If the procedures break down before this theoretical limit on maximum size of unit is reached, it will be described as *not gen app*.

(ii) Non-Rotating

In large scale field surveys it is often desirable to be able to rotate the sample, that is, to drop a portion of the sample and replace it by another at predetermined intervals. The principal reason for wishing to do this is to avoid the kinds of response bias and non-representativeness which can result from being in sample on a number of occasions; phenomena known generally as *sample fatigue*. Rotation will be considered in greater detail in Chapter 3. Meanwhile it should be noted that certain procedures make specific allowance for rotation; that others can be used in rotating samples by selecting initially more units than are required immediately and rotating the excess into the sample as required; and that the remainder (including all those limited to $n = 2$) can only be rotated as whole stratum at a time. Both the latter categories will be described as *non-rotg*.

2.2 DESCRIPTIONS OF SELECTION PROCEDURES

A description of each of the selection procedures listed in Chapter 1 follows.

The format of these descriptions will be

Procedure number and mnemonic title,
Descriptive abbreviations and disadvantages,
Principal references,
Prose description of selection procedure,
Comments.

PROCEDURE 1: Ordered Systematic Procedure

Strπps, *strwor*, *n fixed*, *syst*, *ord*, *b est var*, *j p enum*

Madow (1949), Hartley (1966), Cassel *et al* (1977, p. 17).

Arrange the population units in any convenient order. Cumulate the measures of size down this order. Divide the total measure of size Z by the required number of units in sample, n, to obtain the *skip interval* Z/n. Choose a *random start*, that is, a random number greater than or equal to zero and less than the skip interval. The first unit selected is that for which the cumulated size measure is the smallest greater than or equal to the random start. The second unit is that for which the cumulated size measure is the samllest greater than or equal to the random start plus the skip interval. In general the $(r+1)$th unit selected is that for which the cumulated size measure is the smallest greater than or equal to the random start plus r times the skip interval.

This is the simplest way of selecting a sample with unequal probabilities without replacement. Because of the ordering process, π_{IJ} will be zero for most pairs I, J. In consequence the Yates-Grundy variance estimator will yield considerable under-estimates of variance. Hartley (1966) sought to overcome this disadvantage by making an assumption about the nature of the population sampled. This assumption is that, for any given unit, the value of the variable being estimated depends on the order in which it appears in the population. The population is therefore divided into quasi-strata, one for each sample unit, and the variance calculated accordingly.

PROCEDURE 2: Random Systematic Procedure

Strπps, *strwor*, *n fixed*, *unord*, *b est var*, *j p enum*

Goodman and Kish (1950), Horvitz and Thompson (1952), Hartley and Rao (1962), Rao (1963b), Raj (1964, 1965), Connor (1966), Hanif (1974), Asok and Sukhatme (1976), Cassel *et al* (1977, p. 17).

This procedure is identical with the Ordered Systematic Procedure 1, except that the population units are listed in random order prior to selection.

For this type of selection procedure Hartley and Rao (1962) have given a formula for π_{IJ}, which is asymptotically correct as $N \to \infty$ under certain conditions (see Chapter 1, Section 1.8).

Connor (1966) gave the exact formula for π_{IJ} for any value n and N for this selection procedure.

The main drawbacks of the systematic procedures are the difficulty of calculating the joint probabilities of inclusion for the purpose of estimating the variance, and the fact that one or more of these joint probabilities is sometimes zero. A simple example of a situation in which one of the π_{IJ} is zero is given by $n = 2$; $N = 5$; $Z_I = 1, 2, 4, 5, 6$.

PROCEDURE 3: Grundy's Systematic Procedure

Strπps, strwor, n fixed, syst, unord, b est var, j p enum

Grundy (1954).

This procedure is a modification of the Random Systematic Procedure 2 which avoids listing all population units in random order. Instead, a single unit is selected with probability proportional to size using a procedure originally devised by Lahiri (1951) (see Procedure 45) and the remaining units are selected systematically using the size of the largest unit in the population Z_{max} (or a somewhat larger round number for convenience) as skip interval. Since it is rather difficult to follow this procedure unless it is spelled out step by step, Grundy's description will be repeated (with the notation slightly altered in order to avoid confusion with symbols used elsewhere in this monograph).

(i) Let α denote either Z_{max}, or, if more convenient a round number slightly larger. (The inequalities $Z_{max} \leq \alpha \leq (Z-Z_{max})/(n-1)$ are the essential conditions on α.)

(ii) Choose a random number r in the range $0 < r < \alpha$ and a random integer s in the range 1 to N. If $r \leq Z_s$ accept unit s as a member of the sample and proceed to (iii); otherwise repeat (ii).

(iii) Choose further integers s_2, s_3 in the range 1 to N, distinct from each other and from s, but otherwise random. Note the sizes of the corresponding sampling units. Each time the cumulative sum $r + Z_{s_2} + Z_{s_3} + \ldots + Z_{s_t}$ first exceeds one of the values $\alpha, 2\alpha, \ldots, (n-1)\alpha$, accept the unit s_t as a member of the sample.

This selection procedure may be represented by the following diagram, which corresponds to $n = 4$.

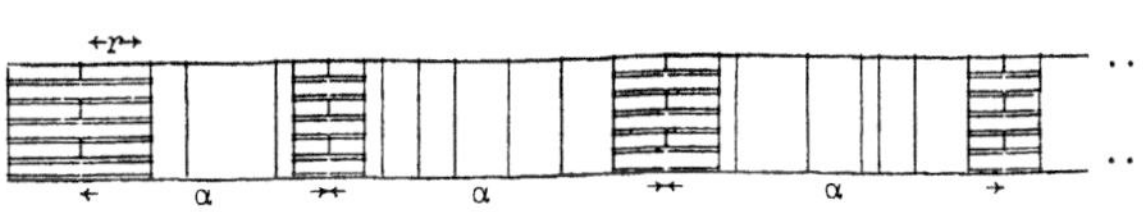

Because the skip interval is smaller than for the Random Systematic Procedure 2, cases where the joint probabilities of selection of one or more pairs of units are zero will be still less common. However it is still relatively easy to produce such cases. One is given by $n = 2$; $N = 5$; $Z_I = 1, 2, 4, 5, 8$; skip interval = 8. The joint probability of selection of the smallest pair is then zero. It is, of course, only under such circumstances that the formulation of an unbiased estimator of variance is impossible.

Another advantage of this modified procedure over the Random Systematic Procedure is that if N is large compared with nZ/Z_{max} it is only necessary to order a small portion of the population randomly.

PROCEDURE 4: Yates Grundy Draw By Draw Procedure

Not *strπps*, *strwor*, *n fixed*, *d by d*, *unord*, *inexact*, *non-rotg*.
Yates and Grundy (1953), Raj (1956a), Murthy (1957), Hanurav (1962b), Hajek (1964), Rao and Bayless (1969), Bayless and Rao (1970), Cassel *et al* (1977, pp. 15, 24, 42-3, 153ff).

Select the first unit in the sample with probability proportional to size Z_I ; the second unit, without replacement, again with probability proportional to size. The total probability of the inclusion of the Ith unit to be in sample is

$$\pi_I = P_I\left(1 + \sum_{J=1}^{N} \frac{P_J}{1-P_J} - \frac{P_I}{1-P_I}\right) . \tag{2.2.1}$$

The joint probability of selecting the Ith and Jth unit is

$$\pi_{IJ} = P_I P_J\left[\frac{1}{1-P_I} + \frac{1}{1-P_J}\right] . \tag{2.2.2}$$

The selection may proceed to $n = 3$ or more, but the formulae for π_I, π_{IJ} and so on, become rapidly complicated.

The procedure is *inexact*, but the unbiased estimators of Raj (1956a) and Murthy (1957), compensate for this inexactness. They will be considered in detail in Chapter 4.

PROCEDURE 5: Yates-Grundy Rejective Procedure

Not *strπps*, *strwor*, *n fixed*, *rej*, *unord*, *inexact*, *non-rotg*.
Yates and Grundy (1953), Durbin (1953a), Hajek (1964).

Select each of the n sampling units in turn with probability proportional to size with replacement. If any unit is selected more than once in the sample, reject the whole sample selected up to that point and continue selecting a new sample with replacement until n distinct units are selected in the sample.

For $n = 2$, the probability of rejecting the initial sample because unit I was selected twice is P_I^2 , that of selecting units I and J in either order is $2P_IP_J$. The probability of including unit I in the final sample is therefore

$$\pi_I = 2P_I(1-P_I) \bigg/ \left(1 - \sum_{\substack{J=1 \\ J\neq I}}^{N} P_J^2\right) . \tag{2.2.3}$$

The sample for $n > 2$ may be obtained similarly, but formulae become complicated very rapidly.

The order of approximation to exactness is poorer than for the Yates-Grundy Draw-by-Draw Procedure 4.

PROCEDURE 6: Midzuno's Procedure

Strπps, strwor, n fixed, d by d, unord, j p iter, not gen app, non-rotg.
Horvitz and Thompson (1952), Yates and Grundy (1953), Rao (1963a).

Select the first unit using a specially calculated set of working probabilities P_I^*, and the remaining units with equal probabilities without replacement. The probabilities used at the first draw are such that the total probability of inclusion of each unit is exactly proportional to size, that is,

$$\pi_I = P_I^* + \frac{n-1}{N-1}\left(1-P_I^*\right) \; , \quad \sum P_I^* = 1 \; . \tag{2.2.4}$$

From (2.2.4),

$$P_I^* = \frac{N-1}{N-n}\pi_I - \frac{n-1}{N-n} \; . \tag{2.2.5}$$

Similarly

$$\pi_{IJ} = \frac{n-1}{N-1}\left[\frac{N-n}{N-2}\left(P_I^*+P_J^*\right) + \frac{n-2}{N-2}\right] \; . \tag{2.2.6}$$

Horvitz and Thompson mentioned that this selection procedure was suggested by Midzuno, presumably on the analogy of his other selection procedure for selecting samples with probabilities proportional to the aggregate measures of size (PPAS) of the units in the sample (Procedure 46). Procedure 6 will break down unless $Z_I \geq Z(n-1)/n(N-1)$ for all I. This is a very stringent requirement; consequently the procedure is frequently not applicable.

Rao (1963a) has shown that for $n = 2$ the variance of y'_{HT} with this Procedure is always smaller than the variance of the *ppswr* estimator y'_{HH} provided $Z_I > Z/2(N-1)$, which is also the condition for non-negativity of the working probabilities.

PROCEDURE 7: Narain's Procedure

Strπps, strwor, n fixed, d by d, unord, $n = 2$ (for all practical purposes), *j p iter, non-rotg.*
Narain (1951), Horvitz and Thompson (1952), Yates and Grundy (1953), Brewer and Undy (1962), Rao (1963b), Cassel *et al* (1977, p. 21).

Calculate a set of working probabilities P_I^*. Select the first unit using these

P_I^* , and the second unit without replacement with probabilities proportional to the same P_I^* . These working probabilities are calculated so that each unit's probability of inclusion in sample is proportional to size. They thus conform to the requirement:

$$\pi_I = P_I^* + \sum_{J \neq I}^{N} \frac{P_I^*}{1-P_J^*} P_J^* . \tag{2.2.7}$$

A numerical procedure for obtaining the P_I^* is given in Appendix A.

The joint probability of inclusion of the Ith and the Jth units together in sample is:

$$\pi_{IJ} = P_I^* P_J^* \left(\frac{1}{1-P_I^*} + \frac{1}{1-P_J^*} \right) . \tag{2.2.8}$$

The sample values of the π_{IJ} are required for the Sen-Yates-Grundy variance estimator.

For $n = 2$ this procedure is generally applicable and the Horvitz and Thompson estimator is always more efficient than the corresponding multinomial sampling estimator. For $n > 2$ the equations for the working probabilities are so complicated that no solution has been proposed.

PROCEDURE 8: Brewer's Procedure

Strπps, strwor, n fixed, d by d, unord, non-rotg, j p enum (for $n > 2$)
Brewer (1963, 1975), Rao and Bayless (1969), Rao and Singh (1973), Chromy (1974), Fuller (1971), Sadasivan and Sharma (1974), Cassel *et al* (1977, p. 16).

Select the rth last sample unit, from among those not already selected, with working probabilities proportional to $P_I(1-P_I)/(1-rP_I)$. Brewer (1975) gives a recursive formula for the joint probabilities of inclusion in sample which involves the consideration of the selection probabilities of all subsets of the sample containing $(n-m)$ units from population of $(N-m)$ $(m = 1, 2, \ldots, n-2)$.

This takes a simple form when $n = 2$, that is,

$$\pi_{IJ} = \left(2P_I P_J \left[\frac{1}{1-2P_I} + \frac{1}{1-2P_J} \right] \right) \div \left(1 + \sum_{K=1}^{N} \frac{P_K}{1-2P_K} \right) . \tag{2.2.9}$$

In this case the joint probabilities of inclusion, and hence also the variance estimator, are simple functions of size. Rao (1963a) found that the Horvitz-Thompson estimator was always more efficient than the corresponding Hansen-Hurwitz estimator for multinomial sampling, and that its variance estimator was never negative. Chromy (1974) found that the π_{IJ} for this procedure (still for $n = 2$) asymptotically minimized the expected variance of the Horvitz-Thompson estimator when $\gamma = \frac{1}{2}$.

PROCEDURE 9: Durbin's 'Method I'

Strπps, strwor, n fixed, d by d, unord, not gen app for $n > 2$.
Durbin (1967), Rao and Bayless (1969), Brewer and Hanif (1970), Fuller (1971), Cassel *et al* (1977, p. 16).

For $n = 2$, select the first unit with probability P_I and the second unit without replacement with probability proportional to $P_J\left[\frac{1}{1-2p_1} + \frac{1}{1-2P_J}\right]$, where p_1 is the sample value of the normed size measure P_I of the unit already selected at the first draw.

The joint probability of inclusion of the Ith and Jth units is

$$\pi_{IJ} = \left\{2P_IP_J\left[\frac{1}{1-2P_I} + \frac{1}{1-2P_J}\right]\right\} \div \left\{1 + \sum_{K=1}^{N} \frac{P_K}{1-2P_K}\right\} \tag{2.2.10}$$

which is the same as given for Procedure 8 (2.2.9). Brewer's Procedure 8 and Durbin's 'Method I' are therefore in the same equivalence class for $n = 2$. Hence in this case Procedure 9 shows the same properties of simplicity of variance estimation and the same superior efficiency vis-à-vis multinomial sampling as Procedure 8.

For $n > 2$, the probability of selection of the Jth population unit at the rth draw, conditional on the results of all the previous draws and in particular given that the Ith unit was selected at the $(r-1)$th draw, is proportional to

$$P_{(r-1)J}\left[\frac{1}{1-2P_{(r-1)I}} + \frac{1}{1-2P_{(r-1)J}}\right] ,$$

where $P_{(r-1)I}$ is the probability of selection of the Ith unit at the $(r-1)$th draw, conditional on the results of the previous draws.

This extension to $n > 2$ is not generally applicable. Fuller (1971) with his 'Scheme A' extended its range of applicability by introducing certain modifications when some units had size Z_I close to Z/n . Even so, the extension remained not generally applicable.

The procedure would be suitable for rotation, in that the probability of inclusion in sample is constant from draw to draw, but for its lack of general applicability for $n > 2$. An example of where the procedure fails to be generally applicable is given by $n = 3$, $N = 102$; $P_1 = P_2 = 0.3$, $P_3 = P_4 = \ldots = P_{102} = 0.004$.

PROCEDURE 10: Durbin's "Grouped Method"

Strπps, strwor, n fixed, d by d, unord, $n = 2$ *only, non-rotg.*

Durbin (1967), Cassel *et al* (1977, p. 16).

Arrange the population units in groups such that each group contains as few units as possible subject to the requirement that the size of the largest unit in each group is less than or equal to half the total size of the group. Select two units from the whole population with replacement. If the units are from different groups, accept both; otherwise accept the first one, replacing the second unit by the second selection using Procedure 9 within the doubly selected group only. For any two units coming from different groups their joint probability of inclusion in sample is $\pi_{IJ} = 2P_I P_J$. For any two units in the same group, their unconditional joint probability of inclusion in sample is

$$\pi_{IJ} = \left[P_I\left(\sum{}' P_I\right)P'_J\left(\frac{1}{1-2P'_I} + \frac{1}{1-2P'_J}\right)\right] \div \left[1 + \sum{}' \frac{P'_K}{1-2P'_K}\right] , \qquad (2.2.11)$$

where $\sum{}'$ denotes summation over the units in the group and

$$P'_I = P_I \div \sum{}' P_I .$$

This selection procedure is slightly less convenient than some others because it requires grouping, but on the other hand it avoids the need for any special calculation of the π_{IJ} if the two units initially selected are from different groups. It thus achieves a measure of simplification in the estimation of variance, but at the cost of some stability in the Sen-Yates-Grundy variance estimator. By using a randomization device in the variance estimation procedure, still further simplification may be obtained at a slight extra cost in stability. This procedure was, in fact, specially devised for handling variance estimation at two or more stages in a simple fashion.

PROCEDURE 11: Rao-Sampford Rejective Procedure

Strπps, *strwor*, n *fixed*, *rej*, *unord*, *non-rotg*.

Rao (1965), Sampford (1967), Rao and Bayless (1969), Bayless and Rao (1970), Asok and Sukhatme (1976), Cassel *et al* (1977, p. 17).

Select the first unit with probability proportional to measure of size. At each subsequent draw select with probability of selection proportional to $P_I/\left(1-nP_I\right)$ with replacement. If any unit is selected twice, reject the whole sample selected and start again. The joint probability of selection for any n is

$$\pi_{IJ} = \frac{I_n\lambda_I\lambda_J}{n(n-1)} \sum_{t=2}^{n} \left[\frac{\left\{t-n\left(P_I+P_J\right)\right\}L_{n-t}(\bar{I},\bar{J})}{n^{t-2}}\right] , \qquad (2.2.12)$$

where

$$K_n = \left[\sum_{I=1}^{N} \frac{tL_{n-t}}{n^t}\right]^{-1} ,$$

$$L_0 = 1 ,$$

$$\lambda_I = Z_I(Z-nZ_I) ,$$

and

$$L_m = \sum_{S(m)} \lambda_{I1}, \lambda_{I2}, \ldots, \lambda_{Im} ,$$

the summation $\sum_{S(m)}$ being over all possible sets of m distinct units drawn from the population. $L_m(\bar{I})$, $L_m(\bar{I}, \bar{J})$ are defined similarly to L_m but relate to the subpopulation formed by omitting unit I and units I and J respectively from the population.

For $n = 2$,

$$\pi_{IJ} = K_2 P_I P_J \left(\frac{1}{1-2P_I} + \frac{1}{1-2P_J}\right)$$

$$= \left\{2P_I P_J \left[\frac{1}{1-2P_I} + \frac{1}{1-2P_J}\right]\right\} \div \left\{1 + \sum_{K=1}^{N} \frac{P_K}{1-2P_K}\right\} \tag{2.2.13}$$

which is identical with (2.2.9) and (2.2.10) so that in this case Procedures 8, 9 and 11 are in the same equivalence class.

This procedure was first suggested by Rao (1965) for $n = 2$ only. Sampford (1967) extended this procedure to cover $n > 2$, but the formula for the π_{IJ} , (2.2.12), is then rather complex. The difficulty in their calculation stems from the large number of decimals which must be stored if they are to be calculated with any acceptable degree of accuracy.

Asok and Sukhatme (1976) compared this procedure with Procedure 2 and proved that the Rao-Sampford Procedure was the more efficient asymptotically for y'_{HT} (1.4.1). They also provided good approximations for the π_{IJ} .

PROCEDURE 12: Durbin-Sampford Procedure

Strπps, strwor, n fixed, d by d, unord, non-rotg.

Sampford (1967), Cassel *et al* (1977, pp. 16-17).

This procedure is difficult to describe for general n . Sampford's description of the procedure for $n = 4$ will therefore be reproduced with appropriate notational changes.

Select the first unit (say the Ith) with probability

$$P_I(1) = C_1\lambda_I \sum_{\substack{K\neq I, J<K \\ I\neq J}}\sum \lambda_J\lambda_K\left(1-P_I-P_J-P_K\right) . \tag{2.2.14}$$

Select the Jth unit if the Ith unit has already been selected with probability

$$P_J(2|I) = C_2\lambda_J \sum_{K\neq I,J} \lambda_K\left(1-P_I-P_J-P_K\right) . \tag{2.2.15}$$

Select the Kth unit when the Ith and Jth units are already selected with probability

$$P_K(3|I, J) = C_3\lambda_K\left(1-P_I-P_J-P_K\right) . \tag{2.2.16}$$

Select the Lth unit when the Ith, Jth and Kth have already been selected with probability

$$P_L(4|I, J, K) = C_4P_L . \tag{2.2.17}$$

Equating the sum of the $P_I(1)$ at each stage to 1 we have

$$\left.\begin{aligned} P_K(3|I, J) &= C_3\lambda_K C_4^{-1} , \\ P_J(2|I) &= C_2\lambda_J C_3^{-1} , \\ P_I(1) &= \tfrac{1}{2}C_1\lambda L C_2^{-1} , \end{aligned}\right\} \tag{2.2.18}$$

so that the probability of drawing the Ith, Jth, Kth and Lth units in that order is $\frac{1}{2}C_1\lambda_I\lambda_J\lambda_K P_L$; that of drawing the Ith, Jth and Kth in any order, followed by the Lth , is $3C_1\lambda_I\lambda_J\lambda_K P_L$ and that of drawing the Ith, Jth, Kth and Lth in any order is

$$3C_1\lambda_I\lambda_J\lambda_K P_K\left(\frac{P_I}{\lambda_I} + \frac{P_J}{\lambda_J} + \frac{P_K}{\lambda_K} + \frac{P_L}{\lambda_L}\right) . \tag{2.2.19}$$

Although this procedure does not set out to use working probabilities, the selection probabilities for the $(n-1)$th and the nth draws are in fact dependent on working probabilities. Consequently for $n = 2$ this procedure is identical in every way with Brewer's Procedure 8. For $n > 2$, the selection probabilities are somewhat more difficult to calculate than $P_I\left(1-nP_I\right)$, so that the Rao-Sampford Procedure 11 is more convenient to use in practice.

PROCEDURE 13: Fellegi's Procedure

Strπps, strwor, n fixed, d by d, unord, j p iter.

Fellegi (1963), Brewer (1967), Rao and Bayless (1969), Bayless and

Rao (1970), Cassel *et al* (1977, p. 16).

Select a unit using working probabilities equal to the normed measures of size. At each subsequent draw select one unit without replacement using working probabilities calculated in such a way that the *a priori* probabilities of selection at that draw are also proportional to size. These working probabilities must be calculated by an iterative procedure, which is fairly simple for the second draw but becomes less tractable as the number of the draw increases, especially if any of the Z_I is close to Z/n. Iterative processes for calculating these working probabilities for $n = 2$ are given in Appendix A. The iteration for $n > 2$ can be slow if the population units are very unequal in size, and care should be taken to ensure that the process has converged before using the working probabilities to select the sample.

This procedure has only been demonstrated to be generally applicable for $n = 2$, but appears to have this property for all values of n.

This procedure was devised specifically to facilitate rotation of the sample. The probability of inclusion in sample is maintained proportional to size of unit because the probability of selecting the Ith unit in sample at each draw is P_I.

PROCEDURE 14: Carroll-Hartley Rejective Procedure

Strπps, strwor, n fixed, rej, unord, j p iter.

Rao (1963b), Carroll and Hartley (1964), Hajek (1964), Rao and Bayless (1969), Bayless and Rao (1970), Cassel *et al* (1977, p. 16).

Select the sample of n units with working probabilities P_I^* with replacement. If not all the units selected are distinct, discard the sample and repeat the same procedure again until n distinct units are selected in the sample. The working probabilities P_I^* must be so chosen that the probability of including the Ith unit in sample is nP_I.

For $n = 2$ this procedure is in the same equivalence class as Fellegi's Procedure 13. For larger values of n it is nearly, but not, quite equivalent (that is, the joint probabilities of selection are nearly the same). Like Fellegi's procedure it appears to be generally applicable for $n > 2$. Rao (1963b) proved that for $n = 2$ the variance of y'_{HT} for this procedure is smaller than that for the corresponding multinomial sampling estimator y'_{HH}.

PROCEDURE 15: Carroll-Hartley Draw-by-Draw Procedure

Strπps, strwor, n fixed, d by d, unord, j p iter.

Carroll and Hartley (1964), Cassel *et al* (1977, p. 16).

This, the draw by draw procedure equivalent of Procedure 14, was used by Carroll and Hartley as an aid in determining the working probabilities for Procedure 14.

To use Procedure 15, as opposed to Procedure 14, involves additional calculation. On the other hand it avoids the selection and consequent rejection of unacceptable samples. For $n = 2$ it is identical in every way with Fellegi's Procedure 13.

PROCEDURE 16: Carroll-Hartley Whole Sample Procedure

Strπps, *Strwor*, *rej*, *n fixed*, *ws*, *unord*, *j p iter*.

Carroll and Hartley (1964), Cassel *et al* (1977, p. 16).

This whole sample equivalent of Procedures 14 and 15 was mentioned by Carroll and Hartley solely for purpose of logical completeness. It is less convenient for selection than either of those procedures and appears to offer no compensating advantages.

PROCEDURE 17: Durbin-Hanurav Procedure (or, Hanurav's Scheme B-A)

Strπps, *strwor*, *n fixed*, *rej*, *unord*, $n = 2$ only.

Durbin (1953b), Hanurav (1966, 1967), Cassel *et al* (1977, p. 16).

(i) Arrange the population units in ascending order of size, so that the normed measure of size of the largest unit is P_N, and of the next largest is P_{N-1}.

(ii) Conduct a Bernoulli trial (Hanurav's Scheme B) in which the probability of success is

$$\delta = \frac{2(1-P_N)(P_N-P_{N-1})}{1-P_N-P_{N-1}}, \quad 0 \le \delta < 1 .$$

(iii) If the trial is successful at step (ii), the sample consists of the largest unit and one other selected from the remainder with probabilities proportional to the measure of size.

(iv) If the trial is not successful at step (ii),

(a) the measure of size of the largest unit is reduced to that of the next largest,

(b) all the measures of size, thus modified, are normed to sum to unity. (These normed measures and sizes will be denoted by P_I^*. Note that $P_N^* = P_{N-1}^*$.)

A sample of two units is then selected using a specifically devised scheme which depends for its validity on the equality of the two largest units. Hanurav has suggested three separate schemes at this point (Schemes A, A' and A"), and each gives rise to a different set of joint probabilities of selection.

(v) The first of these is the scheme A, described below. Select two units with replacement with working probabilities P_I^*. If the sample consists of two distinct units, accept it. Otherwise select two units with replacement with probabilities proportional to P_I^{*2}. Again if the sample consists of two distinct units accept it. Otherwise reject the sample and select two units with replacement with probabilities proportional to P_I^{*4}, and so on, until at the kth trial two distinct units are selected using probabilities proportional to $P_I^{*2^{k-1}}$. The process terminates with probability one.

Durbin (1953b) derived the formulae appropriate to Hanurav's Scheme A, but without Scheme B it could only be used when $P_N = P_{N-1}$. Durbin did not publish this derivation and Hanurav developed Scheme B-A independently of Durbin's work.

Hanurav indicated in an abstract (1966) that this procedure could be extended to $n > 2$. Although full details were not given it would appear that the first part of the procedure would be identical with Steps (i)-(iv) of the Hanurav-Vijayan Procedure 19 (q.v.). The remaining part appropriate to the situation when the population is arranged in ascending order of size and $P_{N-n+1}^* = P_N^*$ is described by Hanurav in his 1966 abstract.

PROCEDURE 18: Hanurav's Scheme B-A'

Strπps, strwor, n fixed, d by d, unord.
Hanurav (1967, 1969), Cassel *et al* (1977, p. 16).

In this selection procedure the first four steps are the same as for the Durbin-Hanurav Procedure 17. The remainder are as follows:

(v) Select the first unit with a set of working probabilities calculated so as to ensure that the total probabilities of inclusion are exactly proportional to size. The working probabilities α_N for this step were given at first incorrectly in Hanurav (1967) but correctly in Hanurav (1969) as

$$\alpha_I = 2P_I^* - \frac{\alpha_1}{N-1} - \frac{\alpha_2}{N-2} - \ldots - \frac{\alpha_{I-1}}{N-I+1}, \quad 1 \le I \le N-1 . \qquad (2.2.20)$$

(Note that when α_N is defined by (2.2.20) it takes the value zero.)

(vi) Select the second unit with equal probabilities from among those units which are later in the population's size ordering then the first selected unit.

PROCEDURE 19: Hanurav-Vijayan Procedure (Hanurav's Scheme B-A")

Strπps, strwor, n fixed, d by d, unord.

Hanurav (1967, 1969), Vijayan (1968), Cassel *et al* (1977, p. 16).

The selection procedure as originally set out by Hanurav is limited to $n = 2$. Steps (i)-(iv) are as for Procedure 17. The remainder comprise Hanurav's Scheme A" and are as follows:

(v) Select the first unit with a special set of working probabilities. These are calculated so as to ensure that the total probabilities of inclusion are exactly proportional to size. The working probabilities β_I for this step were incorrectly given in Hanurav (1967) but correctly in Hanurav (1969) as

$$\beta_I = P_I^*\left\{2 - \frac{\beta_1}{1-P_1^*} - \frac{\beta_2}{1-P_1^*-P_2^*} - \cdots - \frac{\beta_{I-1}}{1-P_1^*-\cdots-P_{I-1}^*}\right\}, \quad 1 \le I \le N-1 . \tag{2.2.21}$$

(Note that when β_N is defined by (2.2.21) it takes the value zero.)

(vi) The second unit is selected with probabilities proportional to P_I^* from among those units later in the population's size ordering then the one selected at step (v).

The selection procedure given by Vijayan for $n \ge 2$ is as follows:

(i) Arrange the population units in ascending order of size.

(ii) Choose one of the numbers 1, 2, ..., n with probabilities δ_i where

$$\delta_i = n\left(P_{N-n+i+1}-P_{N-n+i}\right)\frac{S+iP_{N-n+1}}{S}, \tag{2.2.22}$$

$$S = \sum_{I=1}^{N-n} P_I ,$$

and, by convention, $P_{N+1} = n^{-1}$. (Note: when $n = 2$, δ_1 is the δ of Hanurav's Scheme B and $\delta_2 = 1 - \delta_1$.)

(iii) If the number i is chosen, the last $(n-i)$ population units form part of the sample and the remaining i are selected in accordance with Steps (iv)-(vi) following.

(iv) Define new normed measures of size $P_I^*(i)$ for the $(N-n+i)$ population units as yet unselected where

$$P_I^*(i) = \begin{cases} \dfrac{P_I}{S+iP_{N-n+1}} & \text{for } I < N-n+1 , \\ \dfrac{P_{N-n+1}}{S+iP_{N-n+1}} & \text{for } N-n+1 \le I \le N-n+i . \end{cases} \quad (2.2.23)$$

(v) Select the first of the remaining i units from the first of the $(N-n+1)$ population units as yet unselected with probabilities proportional to $a_j(1)$ where

$$a_1(1) = nP_1^*(i) ,$$

$$a_j(1) = nP_j^*(i) \prod_{k=1}^{j-1} \{1-(i-1)P_k^{**}(i)\} \quad (j = 2, \ldots, N-n+1) ,$$

and where

$$P_k^{**}(i) = \left(P_k^*(i)\right) / \left(\sum_{l=k+1}^{N-n+i} P_l^*(i)\right) . \quad (2.2.24)$$

(vi) If the unit selected at step (v) is the j_1th , select the second unit from the set consisting of the (j_1+1)th up to the $(N-n+2)$nd with probabilities proportional to $a_j(2, j_1)$ where

$$a_{j_1+1}(2, j_1) = (i-1)P_{j_1+1}^*(i) ,$$

$$a_j(2, j_1) = (i-1)P_j^*(i) \prod_{k=j_1+1}^{j-1} \{1-(i-2)P_k^{**}(i)\} ,$$

$$(j = j_1+2, \ldots, N-n+2) . \quad (2.2.25)$$

(vi) Proceed in a similar manner until the last sample unit is selected. In general if the $(l-1)$st of the i units remaining to be selected at the end of step (iv) is chosen to be the j_{l-1}th population unit, then select the lth unit from among the $(j_{l-1}+1)$st up to the $(N-n+l)$th with probabilities proportional to $a_j(l, j_{l-1})$ where

$$a_{j_{l-1}+1}(l, j_{l-1}) = (n-l+1)P_{j_{l-1}+1}^*(i) ,$$

$$a_j(l, j_{l-1}) = (n-l+1)P_j^*(i) \prod_{k=j_{l-1}+1}^{j-1} \{1-(i-l)P_k^{**}(i)\}$$

$$(j = j_{l-1}+1, \ldots, N-n+l) . \quad (2.2.26)$$

PROCEDURE 20: Raj's Variance Minimization Procedure

Strπps, strwor, n fixed, ws, unord, n = 2 only, b est var, non-rotg.
Raj (1956b).

Use a linear programming technique to find those value of π_{IJ} which minimize the variance (1.4.5) of the Horvitz-Thompson estimator (1.4.1) on the assumption that the estimand variables Y_I and measures of size Z_I are related by

$$Y_I = \alpha + \beta Z_I \ , \tag{2.2.27}$$

where α and β are unknown parameters.

This procedure is insufficiently defined for $n > 2$. However values of π_{IJ} determined using the same linear programming technique and minimization criterion can be fed into one of Sinha's Procedures 42-43 (q.v.) to obtain a solution.

The value of the procedure is questionable in that, if the Y_I and Z_I satisfy (2.2.27) exactly, the regression estimator will have zero variance while the Horvitz-Thompson estimator in general, will not. Further, linear programming tends to give extreme solutions with one or more of the π_{IJ} taking the value zero. Hence the Sen-Yates-Grundy variance estimator is seldom, if ever, unbiased.

PROCEDURE 21: Hanurav's Simple Junctional Procedure

Strπps, strwor, n not fixed, d by d, unord[1]*, inexact, b est var, non-rotg.*
Hanurav (1962a), Cassel *et al* (1977, p. 16).

If k_1 is the positive integer such that

$$\delta_1 = \sum_{I=1}^{k_1} \pi_I \le 1 < \delta_1 + \pi_{k_1+1} \ , \tag{2.2.28}$$

select the Ith unit with the probabilities

$$P_I(1) = \begin{cases} \pi_I & \text{if } 1 \le I \le k_1 \ , \\ 1-\delta_1 & \text{if } I = k_1 + 1 \ , \\ 0 & \text{if } I \ge k_1 + 2 \ . \end{cases} \tag{2.2.29}$$

Then (k_1+1)th unit is called a junctional unit. Since it is selected with probabilities $1 - \delta_1$ at the first draw, it will have probability π_{k_1+1} of being selected at least once (and hence of being included in the sample) if its probability

---

[1] Although Hanurav's suggestion is that the population should be arranged in size order, the procedure may be applied with any ordering.

of selection at the second draw is

$$P_{k_1+1}(2) = \left(\pi_{k_1+1} - (1-\delta_1)\right)/\delta_1 \ , \tag{2.2.30}$$

and if at all subsequent draws its probability of selection is zero. The second draw is therefore made with the following selection probabilities:

$$P_I(2) = \begin{cases} 0 & \text{if } 1 \le I \le k_1 \ , \\ \left(\pi_{k_1+1} - (1-\delta_1)\right)/\delta_1 & \text{if } I = k_1 + 1 \ , \\ \pi_I & \text{if } k_1+2 \le I \le k_1+k_2 \ , \\ 1-\delta_2 & \text{if } I = k_1 \ne k_2+1 \ , \\ 0 & \text{if } I \ge k_1+k_2+2 \ , \end{cases} \tag{2.2.31}$$

where k_2 is the integer such that

$$\delta_2 = P_{k_1+1}(2) + \sum_{k_1+2}^{k_1+k_2} \pi_I \le 1 < \delta_2 + \pi_{k_1+k_2+1} \ . \tag{2.2.32}$$

The number of distinct units selected in the first two draws is thus two, with probability $1 - P_{k_1+1}(1)P_{k_1+1}(2)$, and otherwise only one. The procedure is continued until it terminates.

Note that if neither the Ith nor the Jth unit is a junctional unit, and there is no junctional unit between them in the ordering, then $\pi_{IJ} = 0$. Hence an unbiased Sen-Yates-Grundy variance estimator cannot be obtained using this procedure. The next three procedures were devised in order to circumvent this problem.

PROCEDURE 22: Hanurav's Modified Junctional Procedure

Strπps, strwor, n not fixed, d by d, unord, inexact, non-rotg.

Hanurav (1962a), Cassel *et al* (1977, p. 16).

Select one unit from the entire population with probability proportional to size and the remaining units by Procedure 21 in such a fashion that the probability of selection at least once in the second or subsequent draws is

$$\delta_I = \pi_I\left(1 - \frac{1}{\nu}\right)\left(1 - \frac{\pi_I}{\nu}\right) \ . \tag{2.2.33}$$

The procedure does ensure that every pair of population units has a non-zero probability of inclusion in sample. However the joint probabilities of most of the pairs are only of order $P_I P_J/\nu$ and the Sen-Yates-Grundy variance estimator is unstable particularly for large values of ν .

PROCEDURE 23: Hanurav's Double Junctional Procedure

Strπps, strwor, n not fixed, d by d, unord, inexact, non-rotg.
Hanurav (1962a), Cassel *et al* (1977, p. 16).

Select two independent subsamples using Procedure 21, both with the same expected sample number, with the probability of inclusion of the Ith unit in each subsample given by

$$\pi_I' = 1 - \left(1-\pi_I\right)^{\frac{1}{2}} . \tag{2.2.34}$$

A unit is included in sample if it appears in either subsample. The total probability of inclusion of the Ith unit is therefore

$$1 - \left(1-\pi_I'\right)^2 = \pi_I . \tag{2.2.35}$$

PROCEDURE 24: Hanurav's Sequential Procedure

Strπps, strwor, n not fixed, d by d, unord, inexact, non-rotg.
Hanurav (1962a), Cassel *et al* (1977, p. 16).

Select the Ith unit at the rth draw with probability

$$P_I(r) = \frac{P_I\left(K_r-K_{r-1}\right)}{1-K_{r-1}P_I} , \tag{2.2.36}$$

when by convention $K_0 = 0$ and

$$K_r = K_{r-1} + \left[\sum_{I=1}^{N} \frac{P_I}{1-K_{r-1}P_I}\right]^{-1} . \tag{2.2.37}$$

This procedure resembles the three preceding ones in that although selection can occur more than once, inclusion in the sample is dependent only on selection at least once. It differs from these procedures in that the probability of inclusion in sample is proportional to size after each draw, so that sampling may be terminated once a desired sample number is obtained. The expected sample number after r draws is K_r . Hanurav supposed that as $r \to \infty$ all units in the population would eventually be included, so that $K_r \to N$. He deduced that after a certain number of draws, the probabilities of selection would turn negative. In fact, however, the limit of K_r as $r \to \infty$ is $P_{\max}^{-1}$. Since this implies that the probability of inclusion of the largest unit tends to unity as $r \to \infty$, this procedure is generally applicable in a rather special and interesting sense.

PROCEDURE 25: Rao-Hartley-Cochran Procedure

Not strπps, strwor, n fixed, d by d, unord, inexact, non-rotg.

Rao, Hartley and Cochran (1962), Stuart (1964), Raj (1966), Hanurav (1967), Rao and Bayless (1969), Bayless and Rao (1970), Singh and Singh (1974), Chotai (1974), Singh and Kishore (1975), Singh and Lal (1978), Cassel *et al* (1977, pp. 15, 153ff).

Divide the population units at random into n groups of N_J units, $J = 1, 2, \ldots, n$, where the N_J are predetermined. Select one unit from each group with probabilities proportional to the normed measures of size within the group.

A special unbiased estimator for use with this selection procedure will be discussed in Chapter 4. Although this estimator is very similar to the Horvitz-Thompson estimator, the two are not identical and their variance estimators are entirely different.

PROCEDURE 26: Stevens' Procedure

Not strπps, strwor, n fixed, d by d, unord, inexact, non-rotg.

Stevens (1958), Kish (1965), Durbin (1967).

Form the population units into groups of not less than n units each, all units within the group being as close in size as possible. Select n groups with replacement with probability proportional to their total measure of size. If the Jth group is selected n_J times, select n_J units from this group with equal probabilities without replacement. Slight variations of this procedure can be found from author to author and even from the same author but the principle involved is the same.

In practice the restriction to groups of at least n units may be relaxed, but then the procedure ceases to be strictly without replacement.

PROCEDURE 27: Poisson Sampling

Strπps, strwor, n not fixed, d by d, unord, inexact.

Hajek (1964), Ogus and Clark (1971), Brewer, Early and Joyce (1972), Brewer, Early and Hanif (1980), Cassel *et al* (1977, p. 17).

Assign a probability of inclusion in sample to each population unit. Conduct a set of N Bernoulli trials, using each of these probabilities in turn, to determine whether or not the corresponding unit is to be included in sample. The sample consists of all the units for which the trials have been successful.

This procedure will be considered in detail in Chapter 4. The variability of the number of units in sample in this procedure is large as compared with other procedures in which the sample number is a random variable. Nevertheless its extreme simplicity may will commend it for large samples.

Poisson sampling is known in forestry as 3-P sampling; see for instance Sethumadhavi and Rajagopalan (1974).

PROCEDURES 28-31: Hajek's Methods I-IV (respectively)

Not πps, *strwor*, *n fixed*, *rej*, *unord*, *inexact*.
Hajek (1964).

These are four approximations to the Carroll-Hartley Rejective Procedure devised by Hajek (1964) and labelled by him I-IV in descending order of accuracy. They differ from the Carroll-Hartley Rejective Procedure only in their working probabilities, which are given below (but not in Hajek's notation).

Procedure 28: Hajek's "Method I"

$$P_I(r) = \left(P_I\left(1-\lambda P_I\right)/\left(1-nP_I\right)\right) \;/\; \left(\sum_{J=1}^{N} P_J\left(1-\lambda P_J\right)/\left(1-nP_J\right)\right) \qquad (2.2.38)$$

where

$$\lambda^{-1} = \sum_{J=1}^{N} P_J\left(1-nP_J\right) + \left(\sum_{J=1}^{N} P_J^2\left(1-nP_J\right)\right) \;/\; \left(\sum_{J=1}^{N} P_J\left(1-nP_J\right)\right) . \qquad (2.2.39)$$

(Note that λ is of the order of unity.)

Procedure 29: Hajek's "Method II"

$$P_I(r) = \left(P_I\left(1-P_I\right)/\left(1-nP_I\right)\right) \;/\; \left(\sum_{J=1}^{N} P_J\left(1-P_J\right)/\left(1-nP_J\right)\right) . \qquad (2.2.40)$$

Procedure 30: Hajek's "Method III"

$$P_I(r) = \left(P_I/\left(1-\overline{n-1}P_I\right)\right) \;/\; \left(\sum_{J=1}^{N} P_J/(1-\overline{n-1})P_J\right) . \qquad (2.2.41)$$

Procedure 31: Hajek's "Method IV"

$$P_I(r) = \left(P_I/\left(1-nP_I\right)\right) \;/\; \left(\sum_{J=1}^{N} P_J/\left(1-nP_J\right)\right) . \qquad (2.2.42)$$

Hajek's "Method I" appears to be a very accurate approximation. In the example Hajek gives, the probabilities of inclusion are exact to within one or two parts in a thousand. It may therefore be useful as an entry into the iterative procedure for the Carroll-Hartley Rejective Procedure 14.

PROCEDURE 32: Deming's Systematic Procedure

Strπ*ps*, *not strwor*, *n fixed*, *syst*, *ord*, *inexact*.
Deming (1960).

Select systematically m subsamples of n/m units using m different random starts and a skip interval of mZ/n .

Variances may be estimated in an unbiased fashion with $(m-1)$ degrees of freedom by comparing the results of the m systematic samples. The larger m is made, however, the greater the departure from sampling without replacement. This procedure was originally suggested in the context of equal probability sampling, but is just as applicable to sampling with unequal probabilities.

PROCEDURE 33: Variance Estimator Optimization Procedure

Strπps, strwor, n fixed, w.s., n = 2, j p iter.

Brewer and Hanif (1969a).

This procedure fixes the joint probabilities of inclusion in sample in such a fashion as to optimize the stability of the Sen-Yates-Grundy variance estimator under the stochastic model introduced in Chapter 1. Only the π_{IJ} can thus be specified so that this procedure by itself is insufficiently defined for $n > 2$. It can in principle be combined with one of Sinha's Procedures, 42, or 43, to arrive at a defined sample for any value of n. The stability of the Sen-Yates-Grundy variance estimator will be considered in detail in Chapter 3.

PROCEDURE 34: Jessen's Method 1

Strπps, strwor, n fixed, w.s., unord, non-rotg.

Jessen (1969).

Choose a decrement D_m which is an aliquot part of every π_I. (If no such number exists the procedure is not applicable. It is simplest if the largest possible value is chosen for D_m.)

Set out a tableau, in which the first row displays the π_I, $I = 1, 2, \ldots, N$.

The first possible sample consists of these n units with the largest value of π_I. Ties are split arbitrarily. Subtract D_m from those n largest values. Enter in the second row of the tableau $(\pi_I - D_m)$ for those units included in the first sample and π_I for those units not in the first sample.

The rth possible sample consists of those n units with the largest values in the rth row of the tableau. Decrease these values by D_m and enter them in the $(r+1)$th row. Repeat the remaining values from the rth row into the $(r+1)$ row. Continue until the process terminates.

There will then be D_m^{-1} possible samples, some of which will in general be duplicated. Select one of these D_m^{-1} samples with equal probabilities.

EXAMPLE (Modified from Jessen, 1969)

Steps and Decrements	Population Unit No.				Interpretation
	1	2	3	4	
Step No. 1 (π_I)	.2	.4	.6	.8	Sample containing units 3 and 4
Decrements	.0	.0	.2	.2	
Step No. 2	.2	.4	.4	.6	Sample containing units 2 and 4
Decrements	.0	.2	.0	.2	
Step No. 3	.2	.2	.4	.4	Sample containing units 3 and 4
Decrements	.0	.0	.2	.2	
Step No. 4	.2	.2	.2	.2	Sample containing units 1 and 2
Decrements	.2	.2	.0	.0	
Step No. 5	.0	.0	.2	.2	Sample containing units 3 and 4
Decrements	.0	.0	.2	.2	
	.0	.0	.0	.0	

Result: Sample contains units 1 and 2 with probability 0.2 ;
2 and 4 with probability 0.2 ;
3 and 4 with probability 0.6 .

N.B. Jessen used the decrement 0.1 and obtained a different sample space. For comments, see Procedure 36, Jessen's Method 3.

PROCEDURE 35: Jessen's Method 2

Strπps, strwor, n fixed, w.s., unord, non-rotg.
Jessen (1969).

This is identical with Procedure 34, except that the decrement varies from step to step. An $(N+1)$th column is needed, with the initial entry unity, to indicate the probability not yet allocated to any sample prior to the rth step. The entries in this column are decremented at every step.

For the mth step the decrement is chosen to be as large as possible subject to two requirements. The first requirement is that, for every unit to be included in the mth sample, the decrement must not exceed the smallest of the entries in the rth row; otherwise one or more entries in the $(m+1)$th row could be negative. The second requirement is that, for every unit not included in the mth sample, the decrement must not exceed the smallest of the differences between the initial π_I and the entry for the Ith unit in the rth row; otherwise the unit corresponding to that entry would be left with an unallocated balance at the end of the process.

The selection of one out of the possible samples is then made with probabilities given by corresponding decrements.

EXAMPLE (Source: Jessen 1969)

Steps and Decrements	Population Unit No.				$(N+1)$th column (Unallocated Probability)	Intrepretation
	1	2	3	4		
Step No. 1 (π_I)	.2	.4	.6	.8	1.0	Sample containing units 3 and 4
Decrements	.0	.0	.6	.6	.6	
Step No. 2	.2	.4	.0	.2	.4	Sample containing units 1 and 2
Decrements	.2	.2	.0	.0	.2	
Step No. 3	.0	.2	.0	.2	.2	Sample containing units 2 and 4
Decrements	.0	.2	.0	.2	.2	
	.0	.0	.0	.0	.0	

Result: Sample contains units 1 and 2 with probability 0.2 ;
2 and 4 with probability 0.2 ;
3 and 4 with probability 0.6 ;

N.B. This sample space is the same as for the example given under Procedure 34. For comments, see Procedure 36, Jessen's Method 3.

PROCEDURE 36: Jessen's Method 3

Strπps, strwor, n fixed, w.s., unord, non-rotg.

Jessen (1969).

Like Procedure 35, this uses a tableau with a probability decrement that varies from row to row, the $(N+1)$th column indicating, at each step, the probability as yet unallocated to any sample. Here, however, a step may relate to more than one sample.

At the mth step, the units are divided into three sets.

Set 1. Those whose entry in the tableau equals the unallocated probability in the $(N+1)$th column. These units may be included in the rth and all subsequent samples.

Set 2. Those whose entry is intermediate between zero and the corresponding entry in the $(N+1)$th column. These units must be included in some but not all of the mth and subsequent samples.

Set 3. Those whose entry is zero at that step. These units cannot be included in any further samples.

For the mth step, the rule is to include in sample with certainty all units in Set 1, and a random selection of the units in Set 2 large enough to make up the required sample size.

If the probability decrement entered in the $(N+1)$th column at the mth step is

D_m , the entries of those units in Set 1 must each also be decremented by D_m , while those of the units in Set 2 must be decremented by $D_m n_m / N_m$ where N_m is the number of units in Set 2 at the mth step, and n_m is the number of units in Set 2 required to make up the difference between the required sample size and the number of units in Set 1.

The decrement D_m is chosen to be as large as possible subject to two requirements. The first requirement is that $D_m n_m / N_m$ must not exceed the smallest entry for any unit in Set 2 (otherwise the entries could become negative). The second is that D_m must not exceed $N_m / (N_m - n_m)$ times the smallest of the differences between the original π_I and the current tableau entry in the Ith column for any unit in Set 2 (otherwise the unit corresponding to that entry would be left with an unallocated balance at the end of the process).

EXAMPLE (Source: Jessen 1969)

Steps and Decrements	Population Unit No.				(N+1)th Column (Unallocated Probability)	Interpretation
	1	2	3	4		
Step No. 1 (π_I)	.2	.4	.6	.8	1.0	Random non-replacement samples of 2 units from units 1, 2, 3 and 4
Decrements	.2	.2	.2	.2	.4	
Step No. 2	.0	.2	.4	.6	.6	Samples including unit 4 and a random selection of one from units 2 and 3
Decrements	.0	.2	.2	.4	.4	
Step No. 3	.0	.0	.2	.2	.2	Sample containing units 3 and 4 only
Decrements	.0	.0	.2	.2	.2	
	.0	.0	.0	.0	.0	

After completing the tableau, a random number is chosen between 0 and 1 . The type of sample to be selected is indicated by the 'unallocated probability' entry in the (N+1)th column next larger than that random number. In the example given above, a random number in the range 0.2 up to but not including 0.6 would indicate that the sample should include unit 4 and a random selection of one from units 2 and 3.

These first three of Jessen's methods are reasonably convenient for selection purposes and the π_{IJ} are readily calculable. However extreme solutions with one or more of the $\pi_{IJ} = 0$ are common for Procedures 34 and 35. The π_{IJ} for Procedure 36 are always strictly positive. In the example given by Jessen they are identical to those obtained by the Random Systematic Procedure 2. These π_{IJ} are not close to the optimum set for stability of the Sen-Yates-Grundy variance estimator (see Section 3.7).

PROCEDURE 37: Jessen's Method 4

Strπps, strwor, n fixed, w.s., n = 2 only, unord, non-rotg.
Jessen (1969).

Approximate the π_{IJ} by $\left\{\pi_I\pi_J-\left(n-\sum\pi_I^2\right)/N(N-1)\right\}$. Use trial and error to adjust these approximations so as to ensure that $\sum_{J\neq I}^{N}\pi_{IJ}=n\pi_I$ for all I. Select a sample of two units using these π_{IJ} to define the sample space.

This procedure approximates equality for the $\left(\pi_I\pi_J-\pi_{IJ}\right)$ for all combinations I, J. Exact equality (which is generally impossible) would simplify the variance of y'_{HT} to

$$V\left(y'_{\mathrm{HT}}\right)=\left(n-\sum_{I=1}^{N}\pi_I^2\right)\left(\left(\sum_{I=1}^{N}\left(\frac{Y_I}{\pi_I}-\frac{Y}{n}\right)^2\right)/(N-1)\right). \tag{2.2.43}$$

There is no necessity to use trial and error, since the variance of the $\left(\pi_I\pi_J-\pi_{IJ}\right)$ could be minimized analytically. The use of trial and error in Jessen's example results in a set of π_{IJ} which are unfavourable for the stability of the Sen-Yates-Grundy variance estimator. Use of the analytical solution in this case yields a solution close to the theoretical optimum for the stability of this variance estimator. Two alternative formulae for the π_{IJ} which would not require the use of trial and error are given in (3.7.2) and (3.7.3).

PROCEDURE 38: Modified Poisson Sampling

Strπps, strwor, n not fixed, d by d, unord, inexact.
Ogus and Clark (1971), Brewer, Early and Hanif (1980).

Select an ordinary Poisson Sample (Procedure 27). If no units are selected in that sample, reselect repeatedly until a non-empty sample is achieved.

Assuming that the probability of inclusion in the sample of the Ith population unit is to be held constant at π_I, the probability of selecting this unit at each ordinary Poisson draw must be $\pi_I\left(1-P_0^*\right)$ where P_0^* is the probability of selecting an empty sample at each attempt. Then $P_0^*=\prod_{I=1}^{N}\left\{1-\pi_I\left(1-P_0^*\right)\right\}$ and its value may be obtained iteratively using the initial value zero.

Modified Poisson sampling was devised to reduce the variability in sample size which obtains with ordinary Poisson sampling, and in particular to ensure that an empty sample is never selected. This procedure will be discussed in detail in Chapter 4.

PROCEDURE 39: Collocated Sampling

Strπps, strwor, n not fixed, d by d, unord, inexact.

Brewer, Early and Joyce (1972), Brewer, Early and Hanif (1980).

Collocated sampling is similar to Poisson sampling, but reduces the variation in sample size by requiring the random variables r_I used in the Bernoulli trials to be uniformly *spaced* instead of uniformly *distributed* over the interval $[0, 1)$. A random ordering L $(L_I = 1, 2, \ldots, N)$ is chosen with equal probabilities, and a random variable θ is also selected from a uniform distribution over the interval $[0, 1)$. The value of r_I is then defined as $(L_I+\theta-1)/N$. This procedure will also be discussed in detail in Chapter 4.

PROCEDURE 40: Das-Mohanty Procedure

Strπps, strwor, n fixed, w.s., unord, non-rotg.

Das and Mohanty (1973).

Form a sample space containing $b = b_1 + b_2$ samples, each of n distinct units. The b_1 samples are to be such that each population unit appears precisely r times in all samples combined, and that for each pair of units there is to be at least one sample where both appear together. In the b_2 samples the Ith population unit is to appear (cZ_I-r) times in all samples combined. Select one sample at random from the complete sample space giving equal probability to each sample. Das and Mohanty supply three schemes for the construction of sample spaces with these characteristics.

The advantage of this procedure lies in the simplicity of calculation of the π_{IJ}. The difficulty lies in the construction of the sample spaces such that the resulting π_{IJ} provide stable variance estimators. The procedure can be used only when the Z_I are integers. When the original Z_I are not integers, they can - to any desired level of accuracy - be replaced by new integer-valued measures of size. If these are large, however, the procedure becomes difficult to manage. The same is true if Z_{max}/Z is close to n^{-1}, as then the required value of c is large.

PROCEDURE 41: Mukhopadhyay's Procedure

Strπps, strwor, n fixed, unord, d by d, non-rotg.

Mukhopadhyay (1972), Sinha (1973).

This procedure is one which enables a sample of any size n to be selected given the π_I and the π_{IJ} only. A description of this procedure is omitted here for the following reasons:

1. it is extremely complicated both to describe and to use;

2. it can be considered as superseded by Sinha's Procedures 42-43.

Readers who wish to consider this procedure in detail are referred to Mukhopadhyay (1972). It will not be considered further in this monograph.

PROCEDURE 42: Sinha's 'Extension" Procedure

Strπps, strwor, n fixed, w.s., unord, non-rotg.
Sinha (1973).

Given any set of non-negative π_I and π_{IJ} which are feasible in the sense that the π_I sum to n, that the π_{IJ} sum over J to $(n-1)\pi_I$, and that no π_{IJ} exceeds $\min(\pi_I, \pi_J)$ or $(\pi_I+\pi_J-1)$, Sinha's Extension Procedure will provide one possible sample space consistent with that feasible set. The procedure is as follows.

Form the sample space for a sample of $N - 2$ $(\geq n)$ units with inclusion probabilities π_I^* and joint inclusion probabilities π_{IJ}^* given by

$$\left.\begin{aligned} \pi_I^* &= \pi_I(N-2)/n , \\ \pi_{IJ}^* &= \pi_{IJ}(N-2)(N-3)/n(n-1) . \end{aligned}\right\} \qquad (2.2.44)$$

The sample space for this sample of $N - 2$ units is defined by

$$Pr\{\text{sample excludes units } I \text{ and } J\} = 1 + \pi_{IJ}^* - \pi_I^* - \pi_J^* .$$

If $n = N - 2$, the procedure terminates at this point. Otherwise, for each possible sample of $N - 2$ in that sample space, calculate the probabilities of selecting each possible subsample of n from the $N - 2$, using *srswor*. Add these probabilities over the sample space for each possible subsample of n units, and select one such subsample using these probabilities.

It will sometimes happen that the set $\{\pi_I^*, \pi_{IJ}^*\}$ is not feasible in the sense stipulated above for the set $\{\pi_I, \pi_{IJ}\}$. In these cases some of the 'possible' samples of $N - 2$ units will have negative probabilities. If, however, the resulting negative probabilities for the *srswor* subsamples of n units are added algebraically, the procedure will still yield a feasible solution.

PROCEDURE 43: Sinha's Reduction Procedure

Strπps, strwor, n fixed, w.s., unord, non-rotg.
Sinha (1973).

Given any set of non-negative π_I and π_{IJ} which is feasible in the sense defined for Procedure 42, Sinha's Reduction Procedure will provide the identical sample space to that obtainable from Procedure 42. The method is as follows.

Form the sample space for a sample of two units with inclusion probabilities π_I^{**} and joint inclusion probabilities π_{IJ}^{**} given by the implicit formulae

$$\left.\begin{aligned}\pi_I &= \frac{n-2}{N-2} + \frac{N-n}{N-2}\pi_I^{**} , \\ \pi_{IJ} &= \frac{(n-2)(n-3)}{(N-2)(N-3)} + \frac{(N-n)(n-2)}{(N-2)(N-3)}\left(\pi_I^{**}+\pi_J^{**}\right) + \left[1 - \frac{2(n-2)}{N-2} + \frac{(n-2)(n-3)}{(N-2)(N-3)}\right]\pi_{IJ}^{**} .\end{aligned}\right\} \quad (2.2.45)$$

If $n = 2$ the above equations result in the trivial solutions $\pi_I^{**} = \pi_I$, $\pi_{IJ}^{**} = \pi_{IJ}$, and a whole sample of two units can be selected with probabilities π_{IJ} in the usual way. For $n > 2$ the probability of selecting any given sample of n units is the sum (over the sample space of samples of two units) of the probabilities of selecting *srswor* the remaining $(n-2)$ units from the complementary set of $N - 2$ units.

As with Procedure 42 it will sometimes happen that the $\{\pi_I^{**}, \pi_{IJ}^{**}\}$ do not form a feasible set. If, however, the resulting negative probabilities are added algebraically the procedure will still yield a feasible solution.

PROCEDURE 44: Chaudhuri's Procedure

Strπps, strwor, n fixed, d by d, unord, not gen app.
Chaudhuri (1976).

Choose any exact *πpswor* procedure generally applicable for $n = 2$. Select the first two units using that procedure, but with working probabilities calculated so as to ensure that the final probabilities of inclusion (after the entire sample is selected) are proportional to size. Select the remaining $n - 2$ units from among those population units not previously selected, using *srswor*.

This procedure is somewhat analogous to Midzuno's (Procedure 6). Whereas Midzuno selects one unit with unequal and the remainder with equal probabilities, Chaudhuri selects two units with unequal and the remainder with equal probabilities. This procedure is applicable when $P_I > (n-1)/n(N-1)$ for all I.

PROCEDURE 45: Lahiri's Procedure

Not strπps, strwor, n fixed, w.s., unord, inexact, non-rotg.
Lahiri (1951), Sankaranarayanan (1969), Rao and Bayless (1969), Bayless and Rao (1970), Vijayan (1975), Cassel *et al* (1977, pp. 120-121, 154ff).

Select a set of n units using *srswor* and find the aggregate size measure of those units. Choose a random number between zero and the sum of the sizes of the n

largest units (or any number greater than this). If this random number exceeds the aggregate size of the *srswor* sample of n units, reject the sample as a whole; otherwise accept it. If the sample is rejected, repeat the process until a sample is accepted. Clearly the probability that a sample will be accepted is proportional to the aggregate measure of size of the sample units and in consequence the conventional ratio estimator is unbiased. This procedure will be considered in detail in Chapter 4.

PROCEDURE 46: Ikeda-Midzuno Procedure

Not strπps, strwor, n fixed, d by d, unord, inexact, non-rotg.
Midzuno (1952), Avadhani and Srivastava (1972), Singh (1975b).

Select r units using *srswor*. Select the Ith unit from the remaining $N - r$ units with probability $P_I + \sum_{j=1}^{r} p_j/(N-r)$. Select the remaining $n - r - 1$ units using *srswor*. The special case of this procedure with $r = 0$ was devised by Ikeda, and the general case by Midzuno. Like Lahiri's Procedure 45, it selects samples with probabilities proportional to their aggregate measures of size and in consequence the conventional ratio estimator is unbiased. Further discussion will be given in Chapter 4.

PROCEDURE 47: Fuller's Scheme B

Strπps, strwor, n fixed, d by d, n = 2 only, not gen app.
Fuller (1971).

Select the first draw with probabilities equal to the normed measure of size P_I , and at the second with probabilities given by

$$P_{J|I} = \tfrac{1}{2}P_J + \tfrac{1}{2}P_I P_J^2\left(\frac{1}{1+D}\right)\left\{\left(1 \Big/ \sum_{K=1}^{N} P_K^2-2P_I^2\right)+\left(1 \Big/ \sum_{K=1}^{N} P_K^2-2P_J^2\right)\right\} , \qquad (2.2.46)$$

where

$$D = \sum_{L=1}^{N} \left\{P_0^2 \Big/ \sum_{K=1}^{N} P_K^2-2P_L^2\right\} . \qquad (2.2.47)$$

This method has joint probabilities of inclusion π_{IJ} which minimize

$$\sum_{\substack{I,J=1\\J>I}}^{N} \left(\frac{\tfrac{1}{2}\pi_I\pi_J-\pi_{IJ}}{\pi_I\pi_J}\right)^2 , \qquad (2.2.48)$$

and in consequence are nearly proportional to $\pi_I\pi_J$. The procedure is applicable only when

$$2P_{\max}^2 < \sum_{I=1}^{N} P_I^2 , \tag{2.2.49}$$

where $P_{\max}$ is the largest of the P_I .

Fuller (1971) indicated how this procedure could in principle be extended to cover $n > 2$ but did not give details. (N.B. Fuller's Scheme A for $n = 2$ is identical with Procedure 9, q.v.)

PROCEDURE 48: Singh's Procedure

Strπps, strwor, n fixed, syst, ordered, j p enum, not gen app.
Singh (1978).

1. Select a sample of size

$$n' = (N+1)/2 \text{ if } N \text{ is odd}$$
$$= N/2 + 1 \text{ if } N \text{ is even.}$$

(A) Select a random number I from 1 to N by a predetermined probability $P(I)$. For N odd, choose $P(I) = \{n'(Z_I+Z_{I-1})/Z\} - 1$. For N even the specification of $P(I)$ is available but cumbersome.

(B) Starting with I select two contiguous units and thereafter $(n'-2)$ units in a circular systematic fashion with skip interval 2 .

2. Select a sample of the required size n by simple random sampling from the n' already selected.

From the form of $P(I)$ with N odd it will be seen that the method can only be applied when $Z_I + Z_{I-1} \geq Z/n'$ for all I . Even with an optimum arrangement of the units (largest, smallest, second largest, second smallest, and so on) it is easy to produce counter-examples which violate this condition. The procedure does, however (like the other systematic methods), have good rotational properties.

PROCEDURE 49: Choudhry's Procedure

Strπps, strwor, n fixed, d by d, unord, j p iter, non-rotg.
Choudhry (1979).

This uses the Yates and Grundy Procedure 4 for all draws except the last, and at the last uses a set of working probabilities such that the total probabilities of inclusion in sample are proportional to size. For $n = 2$ the procedure is equivalent to Fellegi's Procedure 13. For $n > 2$ it has the advantage that only one set of working probabilities need be calculated instead of $(n-1)$. The procedure appears to be generally applicable for $n > 2$ but no proof of this is available.

PROCEDURE 50: Chromy's Procedure

Strπps, strwor, n fixed, ord, j p enum, non-rotg.

Chromy (1979)

Each population unit is considered in turn, and given a probability of inclusion in sample conditional on the history of the selection process up to that point.

Let $\mathrm{Int}(I) = \left[\sum_{J=1}^{I} \pi_J\right]$ (that is the integral portion of the expression in square brackets) and $\mathrm{Frac}(I) = \sum_{J=1}^{I} \pi_J - \mathrm{Int}(I)$ (that is the fractional portion of the same expression).

By convention $\mathrm{Int}(0) = \mathrm{Frac}(0) = 0$.

The procedure is such that the number of units selected in sample prior to consideration of the Ith population unit is either $\mathrm{Int}(I-1)$ or $\mathrm{Int}(I-1) + 1$. The following table indicates the conditional probabilities of inclusion in sample of the Ith population unit, given the number of units already selected and the relationship between $\mathrm{Frac}(I)$ and $\mathrm{Frac}(I-1)$.

Case No.	Relationship	Conditional probability of inclusion in sample given: $\mathrm{Int}(I-1)$ units previously selected	$\mathrm{Int}(I-1) + 1$ units previously selected
(1)	$\mathrm{Frac}(I) = 0$	1	0
(2)	$\mathrm{Frac}(I) > \mathrm{Frac}(I-1) \geq 0$	$\dfrac{\mathrm{Frac}(I)-\mathrm{Frac}(I-1)}{1-\mathrm{Frac}(I-1)}$	0
(3)	$\mathrm{Frac}(I-1) > \mathrm{Frac}(I) > 0$	1	$\dfrac{\mathrm{Frac}(I)}{\mathrm{Frac}(I-1)}$

(Note that the above table is appropriate only when $0 < \pi_1 < 1$ for all I .)

This selection procedure ensures that at every point in the selection process the expected value of the (cumulated) number of sample units already selected is equal to $\sum_{J=1}^{I} \pi_J$.

It can easily be generalized to the case where some units are so large that $nZ_I/Z > 1$. The number of times such large units are to be included in sample is taken to be $[nZ_I/Z] + 1$ with probability $nZ_I/Z - [nZ_I/Z]$, and $[nZ_I/Z]$ with probability $1 - nZ_I/Z + [nZ_I/Z]$. It is then convenient to express the selection rules in a slightly different kind of table.

Case No.	Relationship	Conditional probability that cumulated number of units selected is to be $\mathrm{Int}(I)+1$ given	
		Previous cumulated number selected was $\mathrm{Int}(I-1)$	Previous cumulated number selected was $\mathrm{Int}(I-1)+1$
(1)	$\mathrm{Frac}(I) = 0$	0	0
(2)	$\mathrm{Frac}(I) \geq \mathrm{Frac}(I-1) \geq 0$	$\dfrac{\mathrm{Frac}(I)-\mathrm{Frac}(I-1)}{1-\mathrm{Frac}(I-1)}$	1
(3)	$\mathrm{Frac}(I-1) > \mathrm{Frac}(I) > 0$	0	$\dfrac{\mathrm{Frac}(I)}{\mathrm{Frac}(I-1)}$

(This table may be used even if some of the nZ_I/Z take zero or integer values.)

To ensure that an unbiased variance estimator can be obtained for samples with $n > 2$, Chromy suggests the following steps:

"(1) Develop an ordered sampling frame of N [population] units;

(2) Select a unit with probability proportional to its size to receive the label 1 ;

(3) Continue labelling serially to the end of the sampling frame;

(4) Assign the next serial label to the first unit at the beginning of the list and continue until all [population] units are labelled;

(5) Apply the sequential ... sample selection algorithm starting with the sampling unit labelled 1 ."

Chromy suggests the use of this selection procedure with meaningfully ordered lists in order to obtain the reductions in variance associated with systematic, stratified or zone sampling. An unordered form of his procedure (starting with a randomized ordering of the population in Step (1) above) could, however, be used, and may yield π_{IJ} values closer to those required for the optimum stability of the Sen-Yates-Grundy variance estimator.

CHAPTER 3

UNEQUAL PROBABILITY PROCEDURES AND THE HORVITZ-THOMPSON ESTIMATOR

3.1. SELECTION PROCEDURES APPROPRIATE FOR USE WITH THE HORVITZ-THOMPSON ESTIMATOR

As mentioned in Chapter 1, the Horvitz-Thompson Estimator has a number of desirable properties when used with an exact sampling procedure. To the three given in Chapter 1 can now be added a fourth, that under model (1.8.5), the expected variance of the Horvitz-Thompson estimator achieves the lower bound of the expected variance for any design-unbiased estimator (Godambe and Joshi, 1965).

The conditions required in Chapter 2 for the description of a procedure as *exact* are that the selection should be strictly without replacement, that the probabilities of inclusion in sample should be strictly proportional to the original measures of size, and that the number of units in sample should be fixed. In Chapter 2, 32 of the 50 selection procedures described had these properties. In this Chapter an attempt will be made to evaluate these procedures under the assumption that they are being used together with the Horvitz-Thompson estimator of total and the Sen-Yates-Grundy estimator of variance.

The criteria for comparison will be limited to samples of size $n = 2$, general applicability, simplicity in selection, simplicity in variance estimation, the efficiency of the Horvitz-Thompson estimator of total, the unbiasedness and stability (that is, efficiency) of the Sen-Yates-Grundy variance estimator, and rotatability.

3.2. LIMITATION TO SAMPLES OF SIZE $n = 2$

If the number of units in the population is large and it is inconvenient to divide into strata, it is imperative that the sample should not be limited to two units. On the other hand it is often convenient to divide large strata into small ones, particularly if they are geographical entities. Moreover, each of the small strata thus formed can usually be made more homogeneous than the original population, making the sample more representative than would be possible without stratification. The case $n = 2$ is, in fact, the limiting case where the maximum advantage in stratification occurs consistent with obtaining an unbiased estimator of variance. (It is possible, by selecting one unit from each of a random subset of the strata and two units from each of the remainder, to push this advantage still further, but it is still necessary to have a suitable method of selecting two units from some of these strata.) Hence the limitation to $n = 2$, while important, is not as critical as it might appear.

The following are formally limited to the case $n = 2$; Procedures 7, 20, 33 and 37. All of these, however, with the exception of Procedure 7, can be extended to the case $n > 2$ by calculating values of π_{IJ} in accordance with the criterion suggested by the chosen procedure, and feeding these π_{IJ} into one of Sinha's Procedures 42-43.

3.3. GENERAL APPLICABILITY

Procedures were described in Chapter 2 as being generally applicable if they could be used given any feasible set of inclusion probabilities $\{\pi_I\}$. To constitute a feasible set the π_I must satisfy the conditions $0 < \pi_I \le 1$ and $\sum_{I=1}^{N} \pi_I = n$ (an integer). Now the desired π_I are typically derived from non-negative size measures Z_I using the relationship $\pi_I = n\, Z_I/Z$ and thus automatically from a feasible set provide only that $Z_{\max} \le Z/n$. It is an obvious and serious inconvenience when feasible sets $\{\pi_I\}$ are encountered which cannot be catered for by a chosen procedure. Those procedures which are not generally applicable even for $n = 2$, namely, Procedures 6, 41, 44, 47 and 48, will therefore be excluded from further consideration. The following procedures, though defined for $n > 2$, are generally applicable for $n = 2$ only; Procedures 9, 10, and 18. These will be considered further in the context of that special case only. Procedures 13, 17, and 49 appear to be generally applicable for $n > 2$, but no proofs are available.

3.4. SIMPLICITY IN SELECTION

Simplicity is of great importance in the choice of a selection procedure, but it is difficult to be entirely objective in the comparison. It is nevertheless possible to draw the reader's attention to the salient features of the various selection processes and provide tentative assessments of their ease or difficulty. This is the approach which will be followed in this Section.

The Systematic Procedures 1, 2, 3, and 48 are particularly simple. Of these, the Ordered Systematic Procedure 1 requires no randomization of the ordering. In Grundy's Systematic Procedure 3, a portion of the population must be put into random order, and in the remaining Procedures 2 and 48, the whole of it. Chromy's Procedure 50 is somewhat less simple than these.

Jessen's two generally applicable decrement-based Procedures 35 and 36, are also quite simple. The decision rules to be used are straigthforward for Procedure 35 but less so for Procedure 36.

For the remaining procedures the cases $n = 2$ and $n > 2$ will be considered separately.

CASE 1: $n = 2$

Sinha's Reduction Procedure 43 takes a particularly simple form when $n = 2$. Since, however, the π_{IJ} are arbitrary, some method must be used for specifying them, which may itself be simple or complicated. Three simple methods will be presented in Section 3.7.

Brewer's Procedure 8 (which for $n = 2$ is identical with Durbin's Method I, that is Procedure 9), the Rao-Sampford Procedure 11, the Durbin-Sampford Procedure 12, the Hanurav Scheme B-A', 18 and the Hanurav-Vijayan Procedure 19 are all quite easy to apply since the Selection Procedure depends on the calculation of probabilities which are simple functions of measure of size. The Rao-Sampford Procedure 11, and the Durbin-Hanurav Procedure 17, being rejective, involve a slight extra complication over the others mentioned. Durbin's Grouped Method (Procedure 10) is less convenient than Durbin's Method I (Procedure 9) in that it requires grouping, but on the other hand it avoids the need for any special calculation whenever the two units initially selected are from different groups. This procedure must also be classed as easy to use.

Jessen's Method 4 (Procedure 37) involves the use of trial and error, and is consequently somewhat inconvenient.

Raj's Variance Minimization Procedure 20 uses linear programming to determine probabilities of whole samples. This is tedious especially when N is large. For any appreciable number of strata a computer program is necessary.

Fellegi's Procedure 13, the Carroll-Hartley Rejective Procedure 14 and the

Carroll-Hartley Draw-by-Draw Procedure 15 involve virtually identical iterative calculations. Several iterative algorithms have been proposed by Fellegi (1963) and by Carroll and Hartley (1964). One of those devised by Carroll and Hartley is claimed to be fairly rapid (2 decimal places per cycle) unless the largest π_I is near unity. Fellegi (1963) also reported quite rapid convergence for his algorithm, provided that all the π_I were less than 0.85-0.90. Alternatively, using the geometrical properties of Fellegi's Procedure pointed out by Brewer (1967) it is possible to obtain an algorithm which achieves a very satisfactory convergence rate even for quite extreme sets of values of π_I. This algorithm, written in BASIC, is given in Appendix A.

Narain's Procedure 7 also requires iteration to obtain the working probabilities used in selection. Procedures for obtaining iterative solutions have been described by Yates and Grundy (1953) and Brewer and Undy (1962). Appendix A contains a BASIC algorithm based on the geometrical properties described by the latter authors. Again a very satisfactory rate of convergence was found even for quite extreme sets of values of π_I.

The approximate values of Hajek's Method I (Procedure 28) may be useful as entry points to iterations for the Carroll-Hartley Rejective Procedure 14.

The Das-Mohanty Procedure 40 is relatively simple when the measures of size can be written as small integers. More usually, however, the number of possible samples which have to be considered is very large, and the selection procedure correspondingly tedious.

The procedures can therefore be arranged in six groups: the Systematic Procedures for which selection is particularly simple; then Jessen's Procedures 35 and 36, together with Sinha's Reduction Procedure 43; next Brewer's Procedure, Durbin's Method I (Procedure 9), Durbin's Grouped Method (Procedure 10), the Rao-Sampford Procedure 11, the Durbin-Sampford Procedure 12, the Durbin-Hanurav Procedure 17, Hanurav's Scheme B-A' 18 and the Hanurav-Vijayan Procedure 19 which are also fairly easy to use; Jessen's Procedure 37, which involves trial and error; Narain's Procedure 7, Fellegi's Procedure 13, the three Carroll-Hartley Procedures 14-16, and Raj's Variance Minimization Proecdure 20, all of which need iterative algorithms; and finally the Das-Mohanty Procedure 40 which in general involves the construction of quite complicated sample space. This last procedure will not be mentioned further.

The Carroll-Hartley Whole Sample Procedure 16 is of course, less simple to use for selection purposes than the corresponding draw-by-draw and rejective procedures. Since it possesses no advantages on the basis of any of the other criteria used in this study, it will also be dropped from the discussion from this point on. Sinha's Extension Procedure 42 is not appropriate for $n = 2$.

CASE 2: $n > 2$

In this situation the Systematic Procedures 1-3 and 48, Chromy's Procedure 50, and Jessen's Procedures 35-36 remain simple. Sinha's Extension Procedure 42 will usually be simple for populations and samples of small size. His Reduction Procedure 43 (which gives the same solution) is generally somewhat less simple for $n > 2$.

Of the remainder, Brewer's Procedure 8 and the Rao-Sampford Procedure 11 retain the same kind of simplicity as for $n = 2$. The Carroll-Hartley Rejective Procedure 14 is again somewhat more convenient to use than its draw-by-draw equivalent Procedure 15. Iteration is required for working out the selection probabilities, both for these for Fellegi's Procedure 13 and for Choudhry's Procedure 49. All the iterative algorithms mentioned for $n = 2$ are available except those based on the geometric properties of the solution. Choudhry's Procedure 49 requires only one set of working probabilities to be calculated iteratively. There is some doubt as to whether the iterative algorithms for Fellegi's Procedure 13 for $n > 2$ converge (see Section 3.7). The approximate working probabilities of Hajek's Method I (Procedure 28) would probably serve as useful entry points into the iterative algorithm for the Carroll-Hartley Rejective Procedure 14.

The order of simplicity is therefore much the same as for $n = 2$; the Systematic Procedures 1-3 and 48 with Jessen's Procedures 34-36, Chromy's Procedure 50 and Sinha's Procedure 42; next Brewer's Procedure 8 and the Rao-Sampford Procedure 11, and finally the Carroll-Hartley Procedures 14-15, Fellegi's Procedure 13, and Choudhry's Procedure 49 all of which require iteration.

3.5. SIMPLICITY IN VARIANCE ESTIMATION

This criterion is closely related to simplicity in selection. This is because the Sen-Yates-Grundy variance estimator contains the joint inclusion probabilities π_{IJ}, and with some notable exceptions (Procedures 1-3 and 48 for all n, Procedures 8 and 11 for $n > 2$) these follow readily from the same kinds of calculations as are needed to carry out selection.

The Ordered Systematic Procedure is in a special position here. Although the π_{IJ} can be simply enough calculated for any given population ordering, many of them will be zero, and the Sen-Yates-Grundy variance estimator will be so biased as to be meaningless. Consequently it is to all intents and purposes impossible to estimate the variance this way.

Connor (1966) produced an exact formula for the π_{IJ} for the Random Systematic Procedure 2. However the evaluation of this formula for any pair of units involves adding contributions from all possible combinations of units separating the two in the pair. This can become tedious for large N. Nevertheless Connor's formula does (at present) make the estimation of variance for the Random Systematic Procedure 2 more amenable to computer programming than it is for the other Systematic Procedures or Chromy's Procedure 50.

CASE 1: $n = 2$

For Sinha's Procedures 42 and 43 the π_{IJ} are arbitrary and a method must be adopted for specifying them. The Rao-Sampford Procedure 11 and the Durbin-Hanurav Procedure 17 involve appreciably more work in estimating variance than in selection. For the Rao-Sampford Procedure 11 the formula for the π_{IJ} is still reasonably compact and involves no iteration. That for the Durbin-Hanurav Procedure 17 is a good deal more difficult to use, because the probabilities of selection change from draw to draw. In this regard it is distinctly less simple to use than Brewer's Procedure 8, Durbin's Method I (Procedure 9), Durbin's Grouped Method (Procedure 10), the Rao-Sampford Procedure 11, the Durbin-Sampford Procedure 12, Hanurav's Scheme B-A' 18, and the Hanurav-Vijayan Procedure 19. It is also easy to use Jessen's Procedures 35-36, as only simple calculations are involved.

Durbin (1967) has suggested the use of a randomization device in the estimation of variance for Durbin's Grouped Method (Procedure 10) which makes it slightly more simple to use than the others. Noting that the value of the coefficient $\left(\pi_1\pi_2\pi_{12}^{-1}-1\right)$ is unity for most pairs of units (see Table 3.1), he suggested that it might be dispensed with entirely by using the value $\left((y_1/\pi_1)-(y_2/\pi_2)\right)^2$ as the estimator with probability $\left\{\pi_1\pi_2\pi_{12}^{-1}-1\right\}$ whenever this value is less than unity, and with probability one when it is equal to or greater than unity. If it exceeds unity for any pair of population units, a bias is introduced, but it would appear to be small for most populations and is actually zero for the case $N = 9$, $n = 2$, $\pi_I = 0.04$, 0.08, 0.08, 0.18, 0.18, 0.24, 0.30, 0.40, 0.50 , as may be seen from an inspection of the bottom left hand triangle in Table 3.1.

A glance at the other half of Table 3.1 suggests that the bias might be serious if Durbin's suggestion were used for the Rao-Sampford Procedure 11, for in this case 16 out of the 36 values of $\left(\pi_1\pi_2\pi_{12}^{-1}-1\right)$ exceed unity. (This device also increases the instability of the variance estimator slightly as will be noted in Section 3.7.) The gain in simplicity therefore amounts to avoiding the calculation for some but not all values of $\left((y_1/\pi_1)-(y_2/\pi_2)\right)^2$. (The use of the above randomization device in multistage sampling will be considered in Chapter 5.)

To summarize: Durbin's Grouped Method (Procedure 10) particularly with randomization device, is slightly easier to use than Brewer's Procedure 8, Durbin's Method I (Procedure 9), the Rao-Sampford Procedure 11, the Durbin-Sampford Procedure 12, Hanurav's Scheme B-A' 18, and the Hanurav-Vijayan Procedure 19, more or less in that order. The Durbin-Hanurav Procedure 17 is next most convenient. All the other relevant procedures, that is Raj's Variance Minimization Procedure 20, Narain's Procedure 7, the three Carroll-Hartley Procedures, Fellegi's Procedure 13, and even more the Random Systematic Procedure 2 and Grundy's Systematic Procedure 3, involve considerable calculations which indicate the need for a computer.

TABLE 3.1

Values of $\left(\pi_I \pi_J \pi_{IJ}^{-1} - 1\right)$ for $N = 9$, $n = 2$;

π_I = 0.04, 0.08, 0.08, 0.18, 0.18, 0.24, 0.30, 0.40, 0.50

Values for Equivalence Class A
(Procedures 8, 9, 11, 12)

Values for Durbin's Grouped Method (Procedure 10)

π_I \ π_J	0.04	0.08	0.08	0.18	0.18	0.24	0.30	0.40	0.50
0.04	-	1.380	1.380	1.240	1.240	1.149	1.051	0.870	0.665
0.08	0.600	-	1.330	1.196	1.196	1.103	1.014	0.840	0.641
0.08	0.600	0.067	-	1.196	1.196	1.108	1.014	0.840	0.641
0.18	1.000	1.000	1.000	-	1.077	0.998	0.913	0.755	0.573
0.18	1.000	1.000	1.000	0.800	-	0.998	0.913	0.755	0.573
0.24	1.000	1.000	1.000	0.200	0.200	-	0.846	0.698	0.528
0.30	1.000	1.000	1.000	1.000	1.000	1.000	-	0.637	0.477
0.40	1.000	1.000	1.000	1.000	1.000	1.000	1.000	-	0.382
0.50	1.000	1.000	1.000	1.000	1.000	1.000	0.250	0.111	-

CASE 2: $n > 2$

Setting aside the procedures considered at the start of the Section, and Sinha's Procedures 42 and 43 for which the π_{IJ} are unspecified, the choice narrows down to Brewer's Procedure 8, the Rao-Sampford Procedure 11, the Durbin-Sampford Procedure 12, Fellegi's Procedure 13 and the Carroll-Hartley Rejective and Draw-by-Draw Procedures 14-15, and Choudhry's Procedure 49. The Rao-Sampford Procedure 11 and the Durbin-Sampford Procedure 12 do not involve iteration and are probably the simplest (though not very simple) for calculating π_{IJ} .

Procedures 8, 13-15 and 49 involve iteration and require a computer. Choudhry's Procedure 49 requires fewer calculations then the remainder of these.

Because of the general complexity of the variance estimation process (especially with $n > 2$) and the relative instability of the Sen-Yates Grundy variance estimator, even when the π_{IJ} are chosen to optimize it, an alternative variance estimator which does not depend on the π_{IJ} will be presented in Section 3.7.

3.6. EFFICIENCY OF THE HORVITZ-THOMPSON ESTIMATOR

Although this might be expected to figure importantly in the comparisons, the efficiency of the Horvitz-Thompson estimator varies so little in practice from one selection procedure to another that it can to all intents and purposes be ignored. Raj (1956b) produced Procedure 20 with a view to minimizing the variance of that estimator, but did so using the assumption that the Y_I and Z_I values were related by the deterministic equation

$$Y_I = \alpha + \beta Z_I \ , \qquad (3.6.1)$$

where α and β were constant. The contribution to the variance from the βZ_I terms is then zero and the variance of y'_{HT} is

$$V(y'_{\mathrm{HT}}) = \alpha^2 V(n'_{\mathrm{HT}}) \ , \qquad (3.6.2)$$

where n'_{HT} is the Horvitz-Thompson estimator of N , obtained by writing unity for each Y_I in (1.4.5).

Raj (1956b) minimized $V(n'_{\mathrm{HT}})$ by linear programming. This characteristically results in an extreme solution in which one or more of the π_{IJ} takes the value zero. As already noted, zero values of π_{IJ} bias the Sen-Yates-Grundy variance estimator. (If a way through this dilemma is sought by setting the relevant π_{IJ} positive but very small, the Sen-Yates-Grundy variance estimator is unbiased but highly unstable.)

Further, model (3.6.1) is only one of many possible derivations from the ideal $Y_I = \beta Z_I$, and other models give different solutions to the problem of variance minimization. In particular if the model is (1.8.5) the expected variance of y'_{HT} , given by Bayless and Rao (1970), is

$$E^*V(y'_{\mathrm{HT}}) = \sigma^2 (Z/n)^{2\gamma} \sum_{I=1}^{N} (1-\pi_I)\pi_I^{2\gamma-1} \ , \qquad (3.6.3)$$

which does not depend on the π_{IJ} at all. Thus under (1.8.5) all exact selection procedures yield equally efficient Horvitz-Thompson estimators.

A similar conclusion may be reached by comparing the asymptotic variance formulae (1.8.1) and (1.8.2). The leading term in these variance formulae is of order N^2. To order N^1 they are identical at

$$V(y'_{\mathrm{HT}}) \doteq \sum_{I=1}^{N} \pi_I\left(1 - \frac{n-1}{n}\pi_I\right)\left((Y_I/\pi_I)-(Y/n)\right)^2 . \qquad (1.8.4)$$

Again the conclusion is that (except for very small populations) the choice of selection procedure has only minimal influence on the efficiency of the Horvitz-Thompson estimator.

There is, however, an exception here in the case of the Ordered Systematic Procedure 1, which is not covered by any such asymptotic variance formulae. The variance of y'_{HT} using this procedure can depend critically on the particular ordering chosen, though not if the Y_I follow the model (1.8.5).

Equations (3.6.3) for the expected variance of y'_{HT} under the model (1.8.5), and (1.8.2) for the asymptotic variance of y'_{HT} to order N^1, may be compared with the corresponding expressions for the Hansen-Hurwitz estimator using multinomial sampling (*ppswr*). These are

$$E^*V(y'_{\mathrm{HH}}) = Z^2 \frac{\sigma^2}{n^{2\gamma}} \sum_{I=1}^{N} \left(1 - \frac{\mu_I}{n}\right)\mu_I^{2\gamma-1} , \qquad (3.6.4)$$

and

$$V(y'_{\mathrm{HH}}) = \sum_{I=1}^{N} \mu_I\left(\frac{Y_I}{\mu_I} - \frac{Y}{n}\right)^2 , \qquad (3.6.5)$$

exactly.

In these expressions, μ_I is written for nP_I, so that μ_I is the expected number of appearances of the Ith population unit in sample. (For without replacement sampling the meanings of π_I and μ_I are identical.) The contribution of the Ith unit to the expected variance is smaller for the y'_{HT} by the factor $(1-\pi_I)/\left(1 - \frac{\mu_I}{n}\right)$. This is very close to the extra factor $\left(1 - \frac{n-1}{n}\pi_I\right)$ which appears in (1.8.4) but not in (3.6.4). Both are of the order of $(N-n)/(N-1)$, the finite population correction factor for simple random sampling.

Rao and Bayless (1969) and Bayless and Rao (1970) conducted both empirical and semi-empirical studies of the efficiency of the Horvitz-Thompson estimator. They compared Procedures 8, 13, and 14 for $n = 2$; Procedures 11, 13, and 14 for $n = 3$; and Procedures 11 and 14 for $n = 4$. They found that with the exception of Procedure 13 for $n = 3$ (which rated as slightly less efficient in the empirical comparison) there were no appreciable differences in the performance of these procedures. The exception would be explained if the iterative algorithm for Procedure 13 failed to converge (see Footnotes 2 and 3 in Bayless and Rao (1970)).

3.7. UNBIASEDNESS AND STABILITY OF THE SEN-YATES-GRUNDY VARIANCE ESTIMATOR

Any procedure which allows any of the π_{IJ} to take the value zero - and this includes Procedures, 1, 2, 3, 20, 34, and 35 - can for that reason result in a biased Sen-Yates-Grundy variance estimator. (It has already been noted that the Ordered Systematic Procedure 1 allows so many of the π_{IJ} to take the value zero that the Sen-Yates-Grundy variance estimator is meaningless.) Sinha's Procedures 42 and 43 provide a special case here, in that the π_{IJ} are arbitrary provided only that they constitute a feasible set.

Rao and Bayless (1969) and Bayless and Rao (1970) used their empirical and semi-empirical populations to compare the stability of the Sen-Yates-Grundy variance estimator for some of the remaining procedures. The semi-empirical studies were based on the model (1.8.5) with normally distributed error terms, under which the expected variance of the Sen-Yates-Grundy variance estimator was taken to be

$$E^*E\{v_{\mathrm{SYG}}(y'_{\mathrm{HT}})-E^*V(y'_{\mathrm{HT}})\}^2 = E^*E\{v_{\mathrm{SYG}}(y'_{\mathrm{HT}})\}^2 - \{E^*V(y'_{\mathrm{HT}})\}^2 ,$$

where

$$E^*E\{v_{\mathrm{SYG}}(y'_{\mathrm{HT}})\}^2 = 3\sigma^4(Z/2)^{4\gamma} \sum_{\substack{I,J=1\\J>I}}^{N}\sum \frac{\pi_I\pi_J-\pi_{IJ}}{\pi_{IJ}} \left(\pi_I^{2\gamma-2}+\pi_J^{2\gamma-2}\right)^2 , \qquad (3.7.1)$$

and $E^*V(y'_{\mathrm{HT}})$ is given by (3.6.3), which is a function of the π_{IJ}. Some procedures which differ quite considerably in the actual process of selection end up with the same set $\{\pi_{IJ}\}$ for any given set $\{\pi_I\}$. The following equivalence classes were recognised for $n = 2$ in Chapter 1.

Equivalence Class A, Procedures 8, 9, 11, 12,

Equivalence Class B, Procedures 13, 14, 15, 16, 49.

For the case $n > 2$ the only equivalence class known is:

Equivalence Class B_1, Procedures 14, 15, 16.

In their 1969 paper, Rao and Bayless considered the case $n = 2$, comparing Procedures

8 and 13 (in Equivalence Classes A and B respectively) and Procedure 17. Bayless and Rao (1970) considered the case $n = 3$, comparing Procedures 11, 13 and 14 and the case $n = 4$ comparing Procedures 11 and 14 only (the last being in Equivalence Class B_1).

Their empirical findings were that for $n = 2$ the three procedures had about equally stable variance estimators. Procedure 17 performed rather better for two of the 20 natural populations, but worse in six of the seven artifical populations. For $n = 3$ (and 4) the stabilities of the three (two) variance estimators were virtually identical.

In their semi-empirical studies they found that, for $n = 2$, Procedure 13 (Equivalence Class B) was consistently more stable than Procedure 8 (Equivalence Class A) but that the gains were small. The stabilities of Procedures 13 and 17 were essentially equal, but Procedure 13 was consistently more stable for $\gamma = 1$. Procedure 17 appeared to be consistently more stable than Procedures 8 and 13 for $\gamma = 0.75$. For $n = 3$, Procedure 13 was found to be consistently less stable than the other two, but there is some doubt as to whether the iterative algorithm for Procedure 13 converged (see Section 3.6).

Brewer and Hanif (1969a) carried out similar semi-empirical studies for the case $N = 4$, $n = 2$, $\pi_I = 0.2, 0.4, 0.6, 0.8$ only. The results, shown in Table 3.2, compare values

$$E^*E\{v_{\mathrm{SYG}}(y'_{\mathrm{HT}}) - E^*V(y'_{\mathrm{HT}})\}^2/\{E^*V(y'_{\mathrm{HT}})\}^2 ,$$

that is to say the relative expected variances of the Sen-Yates-Grundy variance estimator, for all the generally applicable exact procedures with the exception of Sinha's Procedures 42-43 for which the π_{IJ} are arbitrary.

In Table 3.2 all the draw-by-draw and rejective procedures can be seen to approximate the Optimization of Stability Procedure 33 for $\gamma = 1$. The values of the π_{IJ} used to construct Table 3.2 together with some others are given in Table 3.3.

Durbin's Grouped Method (Procedure 10) could not be compared with the other procedures for so small population. Table 3.5 shows the comparison between Equivalence Class A and Procedure 10 both without and with the randomization device mentioned in Section 3.4.

For most of the range $0.50 \leq \gamma \leq 1$, the Rao-Sampford Procedure 12 (together with the other procedures in the Equivalence Class A) has a more stable variance estimator than Durbin's Grouped Method (Procedure 10). The reason for this seems to be that the values of $\left(\pi_I\pi_J\pi_{IJ}^{-1} - 1\right)$ for Equivalence Class A decrease with the π_I and π_J, particularly the larger of the pair, in much the same way as the values of this

TABLE 3.2

Values of relative expected variances of Sen-Yates-Grundy variance estimators for $N = 4$, $n = 2$, $\pi_I = 0.2, 0.4, 0.6, 0.8$

Procedure or Equivalence Class	Relative Expected Variances of $v_{SYG}(y'_{HT})$		
	$\gamma = 0.50$	$\gamma = 0.75$	$\gamma = 1$
Optimization of Stability (Procedure 33) for			
$\gamma = 0.50$	6.44	5.96	5.63
$\gamma = 0.75$	6.58	5.86	5.33
$\gamma = 1.00$	7.02	5.98	5.21
Random systematic 2*	10.35	10.65	10.96
Grundy's systematic 3	10.91	8.55	6.85
Equivalence Class A	8.17	6.61	5.43
Equivalence Class B	7.50	6.22	5.27
Narain 7	7.60	6.27	5.29
Durbin-Hanurav 17	7.53	6.24	5.28
Hanurav's Scheme B-A' 18	7.32	6.33	5.64
Hanurav-Vijayan 19	7.43	6.19	5.27

* For this example Jessen's Procedure 36 is equivalent to Procedure 2.

TABLE 3.3

Values of joint probability of inclusion in sample of pairs of units for $N = 4$, $n = 2$, $\pi_I = 0.2, 0.4, 0.6, 0.8$

Procedure or Equivalence Class	Joint Probabilities of Inclusion					
	$I = 1$, $J = 2$	$I = 1$, $J = 3$	$I = 1$, $J = 4$	$I = 2$, $J = 3$	$I = 2$, $J = 4$	$I = 3$, $J = 4$
Optimization of Stability (Procedure 33) for						
$\gamma = 0.50$	0.0422	0.0588	0.0990	0.0990	0.2588	0.4422
$\gamma = 0.75$	0.0386	0.0559	0.1055	0.1055	0.2559	0.4386
$\gamma = 1.00$	0.0344	0.0535	0.1121	0.1121	0.2535	0.4344
Random systematic 2	0.0667	0.0667	0.0667	0.0667	0.2667	0.4667
Grundy's systematic 3	0.0333	0.0333	0.1333	0.1333	0.2333	0.4333
Equivalence Class A	0.0277	0.0535	0.1188	0.1188	0.2535	0.4277
Equivalence Class B	0.0311	0.0530	0.1158	0.1158	0.2530	0.4311
Narain 7	0.0306	0.0531	0.1163	0.1163	0.2531	0.4306
Durbin-Hanurav 17	0.0323	0.0505	0.1172	0.1172	0.2505	0.4323
Hanurav's Scheme B-A'	0.0444	0.0444	0.1111	0.1111	0.2444	0.4444
Hanurav-Vijayan 19	0.0333	0.0500	0.1167	0.1167	0.2500	0.4333
Chromy 50*	0.0480	0.0987	0.0533	0.0533	0.2987	0.4480

* With the population units arranged in ascending order of size for step (1).

TABLE 3.4

Values of relative expected variances of the Sen-Yates-Grundy variance estimator for

$N = 9$, $n = 2$, $\pi_I = 0.04, 0.08, 0.08, 0.18, 0.18, 0.24, 0.30, 0.40, 0.50$

assuming normality of the ε_I

Procedure or Equivalence Class	Related expected variance of $v_{SYG}(y'_{HT})$		
	$\gamma = 0.50$	$\gamma = 0.75$	$\gamma = 1$
Equivalence Class A	4.74	3.01	2.31
Durbin's Grouped Method (Procedure 10) (without randomization device)	4.63	3.36	3.04
Durbin's Grouped Method (Procedure 10) (with randomization device)	4.74	3.46	3.16

TABLE 3.5

Values of π_{IJ} for $N = 9$, $n = 2$;

$\pi_I = 0.04, 0.08, 0.08, 0.18, 0.18, 0.24, 0.30, 0.40, 0.50$

Values for Equivalence Class A

Values for Durbin's Grouped Method (Procedure 10)

π_I \ π_J	0.04	0.08	0.08	0.18	0.18	0.24	0.30	0.40	0.50
0.04	-	.0013	.0013	.0032	.0032	.0045	.0059	.0086	.0120
0.08	.0020	-	.0027	.0066	.0066	.0041	.0119	.0174	.0244
0.08	.0020	.0060	-	.0066	.0066	.0091	.0119	.0174	.0244
0.18	.0036	.0072	.0072	-	.0156	.0216	.0282	.0210	.0572
0.18	.0036	.0072	.0072	.0180	-	.0216	.0282	.0410	.0572
0.24	.0048	.0096	.0096	.0360	.0360	-	.0390	.0565	.0786
0.30	.0060	.0120	.0120	.0270	.0270	.0360	-	.0733	.1026
0.40	.0080	.0160	.0160	.0360	.0360	.0290	.0600	-	.1447
0.50	.0100	.0200	.0200	.0450	.0450	.0600	.1200	.1800	-

coefficient did in the earlier example, whereas for Durbin's Grouped Method (Procedure 10) they are fixed at unity for most pairs of units. This difference is shown up in Table 3.5, in which for convenience the values for Equivalence Class A have been entered above the main diagonal and those for Durbin's Grouped Method (Procedure 10) below it.

The π_{IJ} used in the construction of Table 3.4 are shown in Table 3.5. The values of the factor $\left(\pi_I\pi_J\pi_{IJ}^{-1}-1\right)$ were given in Table 3.1.

Sinha's Procedures 42 and 43 are atypical in that the π_{IJ} are not products of the selection process, but arbitrary inputs. One simple way of choosing a feasible set of π_{IJ} as an input to Sinha's Procedures is to set

$$\left.\begin{aligned} \pi_{IJ} &= A\pi_I\pi_J + B\left(\pi_I+\pi_J\right) + C\left(\pi_I^2+\pi_J^2\right) , \\ &\text{where} \\ A &= \left(n^2\right)/\left(n^2 - \sum_{J=1}^{N} \pi_J^2\right) , \\ B &= -\left(n \sum_{J=1}^{N} \pi_J^2\right) / \left\{\left(n^2 - \sum_{J=1}^{N} \pi_J^2\right)(N-2)\right\} , \\ &\text{and} \\ C &= \left(n^2\right)/\left\{\left(n^2 - \sum_{J=1}^{N} \pi_J^2\right)(N-2)\right\} . \end{aligned}\right\} \quad (3.7.2)$$

These π_{IJ} will not always be non-negative, but for medium to large values of N approach proportionality to $\pi_I\pi_J$. When this proportionality holds, all the factors $\left(\pi_I\pi_J\pi_{IJ}^{-1}-1\right)$ are equal, and this corresponds to a reasonably, though not optimally, stable Sen-Yates-Grundy variance estimator. Departures from optimality can, however, be serious when N is small, and for the case $N = 4$, $n = 2$, $\pi_I = 0.2, 0.4, 0.6, 0.8$ the value of π_{12} is actually zero.

A more generally satisfactory set of values of π_{IJ} is given by the formula

$$\begin{aligned} \pi_{IJ} &= \frac{n-1}{n}\left\{\pi_I\pi_J + \left(\pi_I^2\pi_J^2\right)/\left(\sum_{K=1}^{N} \pi_K^2\right) + \left(\pi_I^4\pi_J^4\right)/\left(\sum_{K=1}^{N} \pi_K^2 \sum_{K=1}^{N} \pi_K^4\right) + \ldots\right\} \\ &= (n-1) \sum_{r=0}^{\infty} \left(\pi_I^{2^r}\pi_J^{2^r}\right) / \left(\prod_{i=0}^{r} \sum_{K=1}^{N} \pi_K^{2^i}\right) . \end{aligned} \quad (3.7.3)$$

This summation converges rapidly even when one or two values of the π_I are close to unity, each term being less than half the preceding one. The resulting π_{IJ} are necessarily positive, and appear to be close to the optimal values when $\gamma = \frac{1}{2}$. For the case $N = 4$, $n = 2$, $\pi_I = 0.2, 0.4, 0.6, 0.8$ this formula yields $\pi_{12} = 0.0427$, $\pi_{13} = 0.0662$, $\pi_{14} = \pi_{23} = 0.0911$, $\pi_{24} = 0.2662$, $\pi_{34} = 0.4427$ (*cf.* Table 3.3). However when two of the π_I are close to unity (3.7.3) may not result in a feasible set of π_{IJ} .

A third possible choice of π_{IJ} can be made as follows. Multiply each of the given π_I values by $2/n$. The scaled down values of π_I then sum to 2 and corresponding π_{IJ} can be calculated using the Brewer-Durbin-Sampford formula (2.2.9). If these are then multiplied up by $n(n-1)/2$, the resulting π_{IJ} are always positive, but again in extreme cases may not constitute a feasible set.

It is of interest to compare the stability of the Sen-Yates-Grundy estimator for sampling *πpswor* with that of the usual variance estimator for multinomial sampling. For the latter, the coefficients of $\{(y_1/\pi_1)-(y_2/\pi_2)\}^2$ are all unity. (In this respect Durbin's Grouped Method (Procedure 10) represents an approach towards multinomial sampling. In view of the explicit use of multinomial sampling as part of this procedure, this result is not surprising.) Using the same assumptions as before, the relative expected variances of the ordinary variance estimator for multinomial sampling are, for this example, 3.98 for $\gamma = \frac{1}{2}$, 2.78 for $\gamma = 3/4$ and 2.57 for $\gamma = 1$. These small values are somewhat illusory, because the variance itself is much larger when sampling is with replacement. To get a more meaningful comparison, the expected variance of the ordinary variance estimator may be divided by $\{E^*V(y'_{HT})\}^2$. This quotient will be referred to as the *comparison relative expected variance* (CREV) of the ordinary multinomial variance estimator. In this example the CREV takes the values 5.13 for $\gamma = \frac{1}{2}$, 3.93 for $\gamma = 3/4$ and 3.91 for $\gamma = 1$. Referring back to Table 3.5 it will be seen that all these values are higher than the corresponding values for both the Rao-Sampford Procedure 11 and Durbin's Group Method 10.

For the earlier example $(N = 4\ ,\ n = 2\ ;\ \pi_I = 0.2, 0.4, 0.6, 0.8)$ the relative expected variances of the ordinary variance estimator for multinomial sampling were 3.81 for $\gamma = \frac{1}{2}$, 3.39 for $\gamma = 3/4$ and 3.29 for $\gamma = 1$. The CREV's, however, were 8.56 for $\gamma = \frac{1}{2}$, 8.90 for $\gamma = 3/4$, and 10.06 for $\gamma = 1$. Comparing these values with those in Table 3.2 we find for this example, as for the other, not only that the expected variance is smaller when the sample is drawn without replacement but also that, provided a draw-by-draw or rejective procedure is used, this smaller expected variance is absolutely (though not relatively) more accurately estimated. This is particularly true for the larger values of γ.

Although in this comparison the Sen-Yates-Grundy variance estimator comes out reasonably well, it is still unstable by comparison with some of the variance estimators used with special estimators of total which will be encountered in Chapter 4.

In view of the difficulties encountered in attempting to evaluate the π_{IJ} (particularly for $n > 2$) and of the relative instability of the Sen-Yates-Grundy variance estimator the following approximate variance estimator may be preferred;

$$v_{\mathrm{APP}}(y'_{\mathrm{HT}}) = \frac{n}{n-1}\left[1 - \left(\sum_{I=1}^{N} \pi_I^{2\gamma}\right)\Big/\left(\sum_{I=1}^{N} \pi_I^{2\gamma-1}\right)\right] \sum_{i=1}^{n} \left((y_i/\pi_i) - (y'_{\mathrm{HT}}/n)\right)^2 , \quad (3.7.4)$$

where γ is chosen to be the best available 'guesstimate' of the parameter γ of model (1.8.5). Fortunately, the value of (3.7.4) is not critically dependent on the value of γ chosen.

The rationale behind this estimator is as follows. Under model (1.8.5),

$$E^*E \frac{n}{n-1} \sum_{i=1}^{n} \left((y_i/\pi_i) - (y'_{\mathrm{HT}}/n)\right)^2 = \sigma^2 (Z/n)^{2\gamma} \sum_{I=1}^{N} \pi_I^{2\gamma-1} . \quad (3.7.5)$$

This is the larger of the two terms in the expected variance of y'_{HT} given in (3.6.3). The ratio of the smaller to the larger term is $\sum_{I=1}^{N} \pi_I^{2\gamma} / \sum_{I=1}^{N} \pi_I^{2\gamma-1}$ and corresponds to the expression n/N found in the finite population correction for equal probability sampling. When $\gamma = \frac{1}{2}$ the ratio is n/N precisely, and although it increases with γ it does not do so rapidly. The manner in which the factor $1 - \sum_{I=1}^{N} \pi_I^{2\gamma} / \sum_{I=1}^{N} \pi_I^{2\gamma-1}$ functions as a finite population correction is further exhibited by remarking that for multinomial sampling the usual unbiased variance estimator may be written

$$v(y'_{\mathrm{HH}}) = \frac{n}{n-1} \sum_{i=1}^{n} \left(\frac{y_i}{\mu_i} - \frac{y'_{\mathrm{HH}}}{n}\right)^2 , \quad (3.7.6)$$

where $\mu_i = n\pi_i$ is the expected number of appearances in sample of the population unit selected at the ith sample draw and corresponds to the π_i of (3.7.4). When γ is completely unknown, the assumption that $\gamma = \frac{1}{2}$ gives the conservative correction factor $(1-n/N)$. For most populations the value of γ is found to lie between 0.6 and 0.9 , and the value 0.75 will usually be a reasonable 'guesstimate'.

3.8. ROTATABILITY

When a number of surveys are to be made at intervals using the same or nearly the same questionnaire, there can be advantages in rotating the sample; that is, in having a regular programme whereby new units are selected to replace old units that have been in the sample for a specified number of surveys.

The advantage of rotation is that the estimate of total can be improved by using information from past periods (Patterson, 1950). A partial overlap between the previous and current samples is required to exploit this improvement. If, however, the aim is to estimate the changes in total between surveys, it is theoretically best to retain an identical sample. Nevertheless even in this case there are practical advantages in rotation. Objections to keeping the sample unchanged include the following.

(i) Respondents from the first few surveys may refuse to co-operate if asked similar questions on too many successive occasions.

(ii) Respondents who took the trouble to give accurate answers in the first few surveys may become careless. They may for instance continue to give the same answers as before, even though their situation has changed. Interviewers may also become careless in a very similar fashion.

(iii) Respondents who reported candidly to a strange interviewer in the first instance may be reluctant to admit a worsening of their situation, especially to an interviewer who is steadily becoming a more familiar figure.

(iv) Respondents who would remain in a given situation if not questioned about it at regular intervals may be stimulated by this questioning to take steps to change it.

These and other related phenomena are known collectively as *sample fatigue*. All of them tend to diminish the representative character of the sample data. Hence it is usual, in repeating surveys, to arrange that portions of the sample be replaced at regular intervals by new sample units, so that none remains in the sample indefinitely.

Rotation is more often important in multistage designs than in single-stage, but can still be treated quite conveniently in terms of single-stage sampling. Fellegi discusses the two possible alternatives for multistage sampling; a third is mentioned here also.

Alternative I is "the exhaustion of the P.S.U.'s (primary sampling units)". This means that a selected unit is replaced when all its available final stage units have been sampled. (In considering the time when the first rotation need be made, the number of "available units" must be determined by a random mechanism. If all final stage units are regarded as "available", rotation introduces a time-dependent bias, initially in favour of large higher stage units.)

Alternative II involves rotating a P.S.U. every rth survey, the P.S.U.'s selected in the most recent n selections constituting the sample. Selected units remain in the sample for a constant period regardless of their measures of size.

Alternative III is a very crude form of rotation (not mentioned by Fellegi) in which the selected samples are retained in a certain number of strata, and reselected in the remainder.

Fellegi's Procedure 13 is particularly appropriate for Alternative II, since the probability of selection of each population unit is proportional to size at every draw. Alternative II may also be used with any other procedure applicable to $n > 2$ in one of two ways.

(a) If it is recognised beforehand that rotation will be necessary, a larger sample than that immediately necessary can be selected, and the order of these sample units randomized. (Randomization is unnecessary if the procedure of selection is symmetric.) The first n in this random order then constitutes the initial sample and rotation proceeds by dropping the first unit for the $(n+1)$st , the second for the $(n+2)$nd , and so on. When the last sample unit has been used, rotation can still proceed in a limited sense by returning to the first unit dropped from the initial sample.

(b) If the initial sample was already selected and used before the need for rotation was recognised, the larger sample described under (a) must first be tested as to whether it contains all the initial sample units. If not, further larger samples must be selected until one is found which does contain all these units. The initial sample is then taken out and its order randomized. The remaining units of the larger sample are also put in random order. Rotation proceeds by dropping the first unit in random order in the initial sample in favour of the first unit in random order in the remainder, and so on. When the remaining units have been exhausted, rotation can only proceed by returning to the first unit dropped from the initial sample.

A limit to the proportion of the population around which rotation is possible, using Alternative II, is provided by the fact that no sample can be selected larger than n_{max} , where n_{max} is the largest integer less than or equal to Z/Z_{max} . Unless n_{max} is appreciably larger than n , this can be a serious limitation.

If rotation of a self-weighting multistage sample is occuring at the lowest stage of sampling, the minimum rate of rotation for the selected higher stage units is set by the period taken to exhaust the smallest such unit selected. This further limits the extent of rotation possible using Alternative II. With Alternative I, rotation can occur around the entire population.

Alternative III is a very crude form of rotation, and does not even guarantee the certain replacement of any sample unit. If the strata in which reselection is to take place are selected randomly, an estimate of the movement in the estimand can be obtained from the remainder. However since this involves the use of strata to represent other strata, this estimate is not likely to be at all accurate.

To sum up, rotatability depends on which alternatives can be used with which methods. The Systematic Procedures 1-3 and 48 are the only ones for which Alternative I is possible. Alternative II can be used very easily with Fellegi's Procedure and the Carroll-Hartley Procedures 14 and 15, and with somewhat more difficulty with any other method valid for $n > 2$. Alternative III is the only possibility for any procedure limited to $n = 2$.

3.9. SUMMARY

In Tables 3.6 and 3.7 a summary of the properties of some procedures compared in this Chapter is given for $n = 2$ and $n > 2$ respectively.

The principal conclusions which may be drawn are as follows:

(a) The Systematic Procedures 2-3 score highly on simplicity of selection and ease of rotation, but relatively poorly on most other counts, particularly those relating to variance estimation.

(b) Within Equivalence Class A, the Rao-Sampford Procedure 11 is particularly good for $n > 2$ or for rotation with $n = 2$. For $n = 2$ and no rotation, a draw-by-draw method (Brewer's Procedure 8, Durbin's Method I (Procedure 9) or the Durbin-Sampford (Procedure 12)) will probably be slightly more convenient.

(c) Durbin's Grouped Method I (Procedure 10) has a slight advantage over the Equivalence Class A procedures for simplicity in variance estimation, but at the cost of some stability in the variance estimator.

(d) The Carroll-Hartley Rejective Procedure 14 is superior to the Rao-Sampford Procedure 11 only in that it is unnecessary to resort to randomization when using Alternative II for rotation. The Rao-Sampford Procedure 11 is simpler both for selection and possibly for variance estimation. Otherwise there is no difference of any importance.

(e) Fellegi's Procedure 13 has the further advantage over the Rao-Sampford Procedure 11 that oversampling is not required for rotation. However, the iterative selection calculations are stated by Carroll and Hartley to be less simple than theirs and may not converge for $n > 2$. Choudhry's Procedure 49 has the advantage that fewer calculations are required. The Rao-Sampford Procedure 11 remains the simplest of the four for selection and possibly also for variance estimation.

(f) The Hanurav-Vijayan Procedure 19 is a reasonable alternative to Equivalence Class A procedures for $n = 2$. The same could perhaps be said for Hanurav's Scheme B-A' 18 also, but the Durbin-Hanurav Procedure 17 is decidedly less simple to handle both for selection and for estimation of variance.

TABLE 3.6

Summary of properties of selection procedures for $n = 2$

Procedures	Type	Equivalence class (if any)	Is $v_{SYG}(y'_{HT})$ unbiased?	Stability of $v_{SYG}(y'_{HT})$	Nature of selection process	Requirement for calculation of $\{\pi_{IJ}\}$	Can alternative I or II be used for rotation?
Random syst 2	*syst*	-	not always	variable	randomization plus systematic	*j p enum*	I
Grundy's syst 3	*syst*	-	not always	variable	partial randomization plus systematic	*j p enum*	I
Narain 7	*d by d*	-	yes	near optimum	iterative algorithm	same iterative algorithm	No
Brewer 8	*d by d*	A	yes	near optimum	simple working probabilities	simple closed formula	II[3]
Durbin (I) 9	*d by d*	A	yes	near optimum	simple conditional probabilities	simple closed formula	No[4]
Durbin (Grouped)	*d by d*	-	yes[1]	fair	grouping plus Procedure 9	simple closed formula	No
Rao-Sampford 11	*rej*	A	yes	near optimum	simple working probabilities	simple closed formula	II[3]
Durbin-Sampford 12	*d by d*	A	yes	near optimum	simple working probabilities	simple closed formula	No
Fellegi 13 Choudhry 49 Carroll-Hartley 14	*d by d*	B	yes	near optimum	iterative algorithm	same iterative algorithm	II
Carroll-Hartley 15	*rej*	B B_1	yes	near optimum	iterative algorithm	iterative algorithm	II[2]

Table 3.6 (continued)

Procedures	Type	Equivalence class (if any)	Is $v_{SYG}(y'_{HT})$ unbiased?	Stability of $v_{SYG}(y'_{HT})$	Nature of selection process	Requirement for calculation $\{\pi_{IJ}\}$	Can alternative I or II be used for rotation?
Durbin-Hanurav 17	*rej*	-	yes	near optimum	simple working probabilities	closed formula	No
Hanurav B-A' 18	*d by d*	-	yes	near optimum	simple working probabilities	simple closed formula	No
Hanurav-Vijayan 19	*d by d*	-	yes	near optimum	simple working probabilities	simple closed formula	No
Raj's variance minimization 20	*w.s.*	-	no	poor	linear programming	linear programming	No
Jessen (2) 35	*w.s.*	-	not always	variable	decremented	simple calculation	No
Jessen (3) 36	*w.s.*	-	not always	variable	decremented	simple calculation	No
Jessen (4) 37	*w.s.*	-	yes	variable	trial and error	trial and error	No
Sinha (Ext) 42	*w.s.*	-	depends on $\{\pi_{IJ}\}$	depends on $\{\pi_{IJ}\}$	examination of possible samples of $N - 2$ units	no calculations required	II[3]
Sinha (Red) 43	*w.s.*	-	depends on $\{\pi_{IJ}\}$	depends on $\{\pi_{IJ}\}$	cumulation of the π_{IJ}	no calculations required	II[3]
Chromy 50	*d by d*	-	yes (?)	variable (?)	sequential	*j p enum*	II[3]

[1] If the randomization device is used to simplify the estimation of variance, $v_{SYG}(y'_{HT})$ will not always be unbiased, and there will be a slightly additional reduction in its stability.
[2] If Alternative II is used, oversampling is needed.
[3] If Alternative II is used, oversampling and random ordering are needed.
[4] If Alternative II is used for rotation with Durbin's Method I it may break down becasue this method is not generally applicable for $n > 2$.

TABLE 3.7

Summary of properties of selection procedures for $n > 2$

Procedures	Type	Equivalence class (if any)	Is $v_{SYG}(y'_{HT})$ unbiased?	Stability of $v_{SYG}(y'_{HT})$	Nature of selection process	Requirement for calculation of $\{\pi_{IJ}\}$	Can Alternative I or II be used for rotation?
Random syst 2	*syst*	-	not always	variable	randomization plus systematic	*j p enum*	I
Grundy's syst 3	*syst*	-	not always	variable	partial randomization plus systematic	*j p enum*	I
Brewer 8	*d by d*	-	yes	near optimum	simple working probabilities	complicated recursive	II[3]
Rao-Sampford 11	*rej*	-	yes	near optimum	simple working probabilities	closed formula	II[3]
Fellegi 13	*d by d*	-	yes	near optimum	iterative algorithm	same iterative algorithm	II
Carroll-Hartley 14	*rej*	B_1	yes	near optimum	iterative algorithm	same iterative algorithm	II[2]
Carroll-Hartley 15	*d by d*	B_1	yes	near optimum	iterative algorithm	same iterative algorithm	II[2]
Jessen (2) 35 Jessen (3) 36	*w.s.*	-	not always	variable	decremented	simple calculation	No
Sinha (Ext) 42	*w.s.*	-	depends on $\{\pi_{IJ}\}$	depends on $\{\pi_{IJ}\}$	examination of possible samples of $N - 2$ units	no calculations required	II[3]
Sinha (Red) 43	*w.s.*	-	depends on $\{\pi_{IJ}\}$	depends on $\{\pi_{IJ}\}$	examination of possible samples of 2 units	no calculations required	II[3]

Table 3.7 (continued)

Procedures	Type	Equivalence class (if any)	Is $v_{SYG}(y'_{HT})$ unbiased?	Stability of $v_{SYG}(y'_{HT})$	Nature of selection process	Requirement for calculation of $\{\pi_{IJ}\}$	Can Alternative I or II be used for rotation?
Choudhry 49	*d by d*	-	yes	near optimum	iterative algorithm	same iterative algorithm	II[3]
Chromy 50	*d by d*	-	yes (?)	variable (?)	sequential	*j p enum*	II[3]

[2] If Alternative II is used, oversampling is needed.

[3] If Alternative II is used, oversampling and random ordering are needed.

(g) The Jessen Procedures 35 and 36 are simple for selection but cannot be rotated easily and score poorly on most other counts.

(h) Sinha's Procedures 42 and 43 look particularly promising for moderate values of n . Since the π_{IJ} are arbitrary they can be chosen to minimize (or using expression (3.7.3) to come close to minimizing) the variance of the Sen-Yates-Grundy variance estimator. For large values of n the procedures become unmanageable.

(j) Because the Systematic Procedures 2 and 3 are so convenient on all counts other than variance estimation, the approximate variance formula (3.7.4) which does not depend on the π_{IJ} may be used to remedy this deficiency.

CHAPTER 4

SELECTION PROCEDURES USING SPECIAL ESTIMATORS

4.1 INTRODUCTION

In Chapter 3 a comparison was made of those selection procedures for which the Horvitz-Thompson estimator possessed the ratio estimator property. It was mentioned, however, in Section 1.7 that certain special estimators had also been devised for use with particular selection procedures, and that in the context of these procedures they also possessed the ratio estimator property. In this Chapter the performance of these special estimators will be compared in the context of their appropriate selection procedures; that is,

(i) Das's estimator with Procedure 4,

(ii) Raj's and Murthy's estimators with Procedure 4,

(iii) the Rao-Hartley-Cochran (RHC) estimator with Procedure 25,

(iv) unbiased and ratio estimators for Poisson sampling with Procedure 27,

(v) unbiased and ratio estimators for Modified Poisson Sampling with Procedure 38,

(vi) unbiased and ratio estimators for Collocated Sampling with Procedure 39, and

(vii) Lahiri's estimator with Procedures 45-46.

4.2 DESCRIPTION OF SPECIAL ESTIMATORS

4.2.1 Das's Estimator

This estimator was devised by Das (1951) for use with Procedure 4, the draw by draw procedure with probabilities proportional to original size at each draw. He first suggested the following ordered linear combinations, which are unbiased estimators of population total Y ,

$$\left.\begin{aligned} t_1' &= y_1/p_1 \\ t_2' &= (1-p_1)y_2/p_1p_2(N-1) \\ &\cdots \\ t_r' &= \left[\prod_{i=1}^{r-1}\left(1-\sum_{j=1}^{k} p_j\right)y_r\right] \div \left[\prod_{i=1}^{r} p_i \prod_{i=1}^{r-1}(N-i)\right] \end{aligned}\right\} . \tag{4.2.1}$$

Every linear combination $t' = \sum_{r=1}^{n} c_r t_r'$, where $\sum_{r=1}^{n} c_r = 1$, is an unbiased estimator of Y . The choice of c_r is free but for simplicity Das chose $c_r = n^{-1}$. The unbiased variance estimator he provided can assume negative values.

Murthy (1957) showed that estimators such as those in (4.2.1) could be improved by *unordering*; that is, taking the expectation of the estimators derived from any given estimator formula by considering all possible orderings (permutations) of the observed sample. The unordered form of t_1' is identical with Murthy's estimator (Pathak 1961). This estimator is considered in Section 4.2.2. Unorderings of t_r' , $r \neq 1$, yield estimators inferior to Murthy's (Samiuddin, Hanif and Asad 1978). These estimators will not be considered further in this monograph.

4.2.2 The Raj and Murthy Estimators

These estimators were devised by Raj (1956a) and Murthy (1957) for use with Procedure 4. The set of unbiased and mutually uncorrelated estimators of population total Y suggested by Raj is

$$\left.\begin{aligned} t_1 &= y_1/p_1 , \\ t_2 &= y_1 + y_2(1-p_1)/p_2 , \\ &\cdots \\ t_n &= y_1 + y_2 + \dots + y_{n-1} + \frac{y_n}{p_n}(1 - p_1 - p_2 - \dots - p_{n-1}) \end{aligned}\right\} . \tag{4.2.2}$$

The estimator t_{mean} of the population total Y is the arithmetic mean of the above set of estimators, which for $n = 2$ yields

$$t_{\text{mean}} = \tfrac{1}{2}\left[(1+p_1)\,\frac{y_1}{p_1} + (1-p_1)\,\frac{y_2}{p_2}\right] , \tag{4.2.3}$$

with variance

$$V(t_{\text{mean}}) = \frac{1}{8} \sum_{\substack{I,J=1 \\ J \neq I}}^{N}\sum P_I P_J (2-P_I-P_J) \left(\frac{Y_I}{P_I} - \frac{Y_J}{P_J}\right)^2 . \tag{4.2.4}$$

An unbiased estimator of (4.2.4) given by Raj is

$$v(t_{\text{mean}}) = \frac{(1-p_1)^2}{4} \left(\frac{y_1}{p_1} - \frac{y_2}{p_2}\right)^2 . \tag{4.2.5}$$

Pathak (1967a) derived a formula for the variance of t_{mean} for any n . This variance formula is

$$V(t_{\text{mean}}) = \frac{1}{2n^2} \sum_{\substack{I,J=1 \\ J \neq I}}^{N}\sum P_I P_J \left[1 + \sum_{r=2}^{n} Q_{IJ}(r-1)\right]\left[\frac{Y_I}{P_I} - \frac{Y_J}{P_J}\right]^2 , \tag{4.2.6}$$

where $Q_{IJ}(r-1)$ denotes the probability of non-inclusion of one or both of the units I and J in the first $(r-1)$ sample units.

An unbiased estimator of variance suggested by Raj (1956a) for any n is

$$v(t_{\text{mean}}) = \frac{1}{n(n-1)} \sum_{k=1}^{n} (t_k-\bar{t})^2 , \tag{4.2.7}$$

which is non negative for all $n \geq 2$. Here $\bar{t} = \frac{1}{n}\sum_{k=1}^{n} t_k$.

Murthy (1957) suggested that the estimator t_{mean} could be improved by the process of unordering. For $n = 2$ the unordered form of t_{mean} , denoted by t_{symm} , may be written as follows:

$$t_{\text{symm}} = \frac{1}{2-p_1-p_2}\left[\frac{y_1}{p_1}(1-p_2) + \frac{y_2}{p_2}(1-p_1)\right] . \tag{4.2.8}$$

The variance of t_{symm} for $n = 2$ is

$$V(t_{\text{symm}}) = \tfrac{1}{2} \sum_{\substack{I,J=1 \\ J \neq I}}^{N}\sum P_I P_J \frac{1-P_I-P_J}{2-P_I-P_J}\left[\frac{Y_I}{P_I} - \frac{Y_J}{P_J}\right]^2 . \tag{4.2.9}$$

An unbiased variance estimator of (4.2.9) is

$$v(t_{\text{symm}}) = \frac{(1-p_1)(1-p_2)(1-p_1-p_2)}{(2-p_1-p_2)^2}\left[\frac{y_1}{p_1} - \frac{y_2}{p_2}\right]^2 . \tag{4.2.10}$$

Murthy (1957) further showed that an unordered and therefore more efficient unbiased variance estimator for t_{mean} for $n = 2$ is

$$v_M(t_{\text{mean}}) = \tfrac{1}{4}(1-p_1)(1-p_2)\left[\frac{y_1}{p_1} - \frac{y_2}{p_2}\right]^2 . \tag{4.2.11}$$

Pathak (1967a) derived the following variance formula for t_{symm} for any $n \geq 2$:

$$V(t_{\text{symm}}) = \tfrac{1}{2} \sum_{\substack{I,J=1 \\ J \neq I}}^{N}\sum P_I P_J \left\{1 - \sum_{s \ni IJ}^{*} \frac{p(s|I)p(s|J)}{p(s)}\right\}\left(\frac{Y_I}{P_I} - \frac{Y_J}{P_J}\right)^2 , \tag{4.2.12}$$

where $p(s)$ denotes the probability of obtaining the sample s of n units, $p(s|I)$ denotes the probability of obtaining the sample s given that unit I was drawn first, and $\sum_{s \ni IJ}^{*}$ denotes the sum over all samples s containing units I and J .

Pathak (1967b) also derived the following unbiased variance estimator for any n :

$$v(t_{\text{symm}}) = \tfrac{1}{2} \sum_{\substack{i,j=1 \\ j \neq i}}^{n}\sum p_i p_j [p(s)p(s|ij)-p(s|i)p(s|j)]p(s)^{-2}\left[\frac{y_i}{p_i} - \frac{y_j}{p_j}\right]^2 , \tag{4.2.13}$$

where $p(s|ij)$ denotes the conditional probability of selecting the observed sample s , given that units i and j were selected in that order at the first two draws.

(4.2.13) is non negative but the computation becomes cumbersome as n increases. Bayless (1968) developed a computer programme to calculate $p(s|ij)$, $p(s|i)$ and $p(s)$ for *upswor*.

Pathak (1961) showed that Murthy's estimator (4.2.8) could be obtained by unordering any linear combination of the individual Raj estimators (4.2.2).

Note. Basu (1970) suggested that it was natural to estimate the ratio

$$\left(\sum_{I=1}^{N} Y_I - \sum_{i=1}^{n} y_i\right) \div \left(\sum_{I=1}^{N} P_I - \sum_{i=1}^{n} p_i\right)$$

by some sort of an average of the observed ratios. Two particular averages which he suggested were $\sum_{i=1}^{n} y_i \div \sum_{i=1}^{n} p_i$, which led to the conventional ratio estimator, and

$n^{-1} \sum_{i=1}^{n} y_i/p_i$ which led to the estimator

$$\hat{y}_B = \sum_{i=1}^{n} y_i + \frac{1}{n} \sum_{i=1}^{n} \frac{y_i}{p_i} \left(1 - \sum_{j=1}^{n} p_j\right) . \quad (4.2.14)$$

He claimed that these two estimators had 'as much face validity' as unordered forms of the individual Raj estimator (4.2.2), and that although they were not unbiased, they were far simpler to calculate. His argument for 'face validity' appears to be based on their being symmetric functions of the sample values and possessing the ratio estimator property. The authors are not aware of any investigation that has been made as to the performance of Basu's estimators with Procedure 4, but they are not design unbiased, even asymptotically.

4.2.3 The Rao-Hartley-Cochran Estimator

The RHC sampling scheme (Procedure 25) has already been described in Chapter 2. The population units are divided randomly into groups containing N_J units, $J = 1, 2, 3, \ldots, n$, where the N_J are predetermined. One unit is selected from each group, the probabilities of selection being the normed measures of size within the group. The RHC estimation procedure is to form the Horvitz-Thompson estimator for each group separately, and add over the groups.

The unbiased estimator of population total Y is, therefore,

$$y'_{\text{RHC}} = \sum_{i=1}^{n} \frac{y_{it}\pi_i}{p_{it}} , \quad (4.2.15)$$

where p_{it} is the sample value of the normal measure of size P_{iT} , $\pi_i = \sum_{T=1}^{N_i} P_{iT}$, and $\sum_{i=1}^{n} \pi_i = 1$.

The variance of (4.2.15) is

$$V(y'_{\text{RHC}}) = \left[n\left(\sum_{i=1}^{n} N_i^2 - N\right) / N(N-1)\right]\left[\sum_{i=1}^{n} \sum_{T=1}^{N_i} \frac{Y_{iT}^2}{nP_{iT}} - \frac{Y^2}{n}\right] . \quad (4.2.16)$$

Rao, Hartley and Cochran minimized (4.2.16) by noting that since $N = nR + k$, where $0 < k < n$ and R is a positive integer, it was possible to put $N_1 = N_2 = \ldots = N_k = R+1$ and $N_{k+1} = N_{k+2} = \ldots = N_n = R$, in which case (4.2.16) reduces to

$$V(y'_{\text{RHC}}) = \left\{1 - \frac{n-1}{N-1} + \frac{k(n-k)}{N(N-1)}\right\}\left\{\sum_{i=1}^{n} \sum_{T=1}^{N_i} \frac{Y_{iT}^2}{nP_{iT}} - \frac{Y^2}{n}\right\} . \quad (4.2.17)$$

If N is a multiple of n , $k = 0$, and the variance further reduces to

$$V(y'_{\text{RHC}}) = \left(1 - \frac{n-1}{N-1}\right)\left(\sum_{i=1}^{n} \sum_{T=1}^{N_i} \frac{Y_{iT}^2}{nP_{iT}} - \frac{Y^2}{n}\right) . \tag{4.2.18}$$

An unbiased variance estimator of (4.2.16) is

$$v(y'_{\text{RHC}}) = \left[\left(\sum_{i=1}^{n} N_i^2 - N\right) / \left(N^2 - \sum_{i=1}^{n} N_i^2\right)\right] \sum_{i=1}^{n} \pi_i \left(\frac{y_{it}}{p_{it}} - y'_{\text{RHC}}\right)^2 . \tag{4.2.19}$$

For the simpler forms (4.2.17) and (4.2.18), (4.2.19) reduces to

$$v(y'_{\text{RHC}}) = \frac{N^2+k(n-k)-Nn}{N^2(n-1)-k(n-k)} \sum_{i=1}^{n} \pi_i \left(\frac{y_{it}}{p_{it}} - y'_{\text{RHC}}\right)^2 \tag{4.2.20}$$

and

$$v(y'_{\text{RHC}}) = \frac{1}{n-1}\left(1 - \frac{n}{N}\right) \sum_{i=1}^{n} \pi_i \left(\frac{y_{it}}{p_{it}} - y'_{\text{RHC}}\right)^2 . \tag{4.2.21}$$

4.2.4 Poisson Sampling

Poisson sampling as defined by Hajek (1964) gives each unit in the population a certain probability of inclusion in the sample which will be denoted by π_I for the Ith unit, $I = 1, 2, \ldots, N$. To select, a set of N binomial trials is carried out to determine whether each unit in turn is to be included in the sample s or not.

The unbiased Horvitz-Thompson estimator of the population total is

$$y'_{\text{PS}} = \sum_{i \in s} \frac{y_i}{\pi_i} . \tag{4.2.22}$$

Since the joint probability of inclusion π_{IJ} takes the simple form $\pi_{IJ} = \pi_I \pi_J$, the variance of (4.2.22) is

$$V(y'_{\text{PS}}) = \sum_{I=1}^{N} (1-\pi_I) \frac{Y_I^2}{\pi_I} , \tag{4.2.23}$$

and an unbiased estimator of (4.2.23) is

$$v(y'_{\text{PS}}) = \sum_{i \in s} (1-\pi_i) \frac{y_i^2}{\pi_i^2} . \tag{4.2.24}$$

Because the sample size varies in this sampling procedure, the ratio estimator

$$y''_{PS} = \begin{cases} \dfrac{y'_{PS}}{m} \cdot n & \text{if } m > 0 \,, \\ 0 & \text{otherwise,} \end{cases} \tag{4.2.25}$$

is more efficient than y'_{PS} .

The mean square error of y''_{PS} is given approximately by

$$V(y''_{PS}) \doteq \sum_{I=1}^{N} \pi_I(1-\pi_I)\left(\frac{Y_I}{\pi_I} - \frac{Y}{n}\right)^2 + P_0 Y^2 \,, \tag{4.2.26}$$

where $P_0 = Pr(m = 0)$ and $n = E(m) = \sum_{I=1}^{N} \pi_I$. (A proof of (4.2.26) is given in Appendix B.)

The conventional estimator of the approximation (4.2.25) is

$$v(y''_{PS}) = \sum_{i\in s} (1-\pi_i)\left(\frac{y_i}{\pi_i} - \frac{y''_{PS}}{n}\right)^2 + P_0 y''^2_{PS} \,; \tag{4.2.27}$$

but a more stable estimator is obtained by multiplying the first expression on the right hand side by n/m .

4.2.5 Modified Poisson Sampling

Modified Poisson sampling is a procedure which ensures that an empty sample is never selected. It was first suggested by Ogus and Clark (1971). An ordinary Poisson sample is drawn first, but if there are no units in that sample, a second Poisson sample is drawn, and so on repeatedly until a non-empty sample is achieved.

Assuming that the probability of including the Ith population unit in sample is to be held constant at π_I , the probability of selecting this unit in each ordinary Poisson sample drawn must be $\pi_I(1-P_0^*)$, where P_0^* is the probability of selecting an empty sample at each such draw. Then

$$P_0^* = \prod_{I=1}^{N} \{1-\pi_I(1-P_0^*)\} \,,$$

and its value may be obtained iteratively using the initial value zero. For modified Poisson sampling

$$\pi_{IJ} = \pi_I\pi_J(1-P_0^*) \,, \text{ for } I \neq J \,.$$

The variance of the Horvitz-Thompson estimator, y'_{MPS} , formed analogously to (4.2.22) but using modified Poisson sampling is

$$V\left(y'_{\text{MPS}}\right) = \sum_{I=1}^{N} \left(1-\pi_I\right) \frac{Y_I^2}{\pi_I} - P_0^* \left\{Y^2 - \sum_{I=1}^{N} Y_I^2\right\} , \qquad (4.2.28)$$

and an unbiased estimator of this variance is

$$v\left(y'_{\text{MPS}}\right) = \sum_{i\in s} \left(1-\pi_i\right) \frac{y_i^2}{\pi_i^2} - \frac{P_0^*}{1-P_0^*} \left\{y'^2_{\text{MPS}} - \sum_{i\in s} \frac{y_i^2}{\pi_i^2}\right\} . \qquad (4.2.29)$$

The mean square error of the ratio estimator y''_{MPS} , formed analogously to (4.2.25) is approximately given by

$$V\left(y''_{\text{MPS}}\right) \doteq \sum_{I=1}^{N} \pi_I\left\{1-\left(1-P_0^*\right)\pi_I\right\} \left(\frac{Y_I}{\pi_I} - \frac{Y}{n}\right)^2 . \qquad (4.2.30)$$

The conventional estimator of this approximate mean square error is

$$v\left(y''_{\text{MPS}}\right) = \sum_{i\in s} \left\{1-\left(1-P_0^*\right)\pi_i\right\} \left(\frac{y_i}{\pi_i} - \frac{y''_{\text{MPS}}}{n}\right)^2 , \qquad (4.2.31)$$

but a more stable estimator can be obtained by multiplying this expression by n/m . We notice that $V\left(y'_{\text{MPS}}\right) < V\left(y'_{\text{PS}}\right)$ and that provided

$$P_0^* \sum_{I=1}^{N} \pi_I \left(\frac{Y_I}{\pi_I} - \frac{Y}{n}\right)^2 < P_0 Y^2 \qquad (4.2.32)$$

- a condition which is easily satisfied - it is also true that $V\left(y''_{\text{MPS}}\right) < V\left(y''_{\text{PS}}\right)$. Despite this, the only advantage of modified Poisson sampling over ordinary Poisson sampling is that it ensures a non-empty sample. If the sample selected is much smaller (or much larger) than the target size, modified Poisson sampling provides no remedy and will therefore receive no further consideration in this monograph. A procedure which ensures a more stable sample size is described in the following Subsection.

4.2.6 Collocated Sampling

Collocated sampling is similar to Poisson sampling, but reduces the variation in sample size by requiring the random variable r_I to be uniformly *spaced* instead of uniformly *distributed* over the interval $[0, 1)$. A random ordering L $\left(L_I = 1, 2, \ldots, N\right)$ is chosen with equal probabilities, and a random variable θ is also selected from a uniform distribution over the interval $[0, 1)$. For each I we then define

$$r_I = \frac{L_I+\theta-1}{N} . \qquad (4.2.33)$$

The Horvitz-Thompson estimator is still used, but now no simplification of its variance formula is possible. The variance of y'_{CS}, formed analogously to (4.2.22) is therefore

$$V(y'_{CS}) = \sum_{I=1}^{N} (1-\pi_I) \frac{Y_I^2}{\pi_I} + 2 \sum\sum_{\substack{I,J=1 \\ J>I}}^{N} (\pi_{IJ}-\pi_I\pi_J) \frac{Y_I Y_J}{\pi_I\pi_J} . \qquad (4.2.34)$$

An unbiased variance estimator is well known to be

$$v(y'_{CS}) = \sum_{i\in s} (1-\pi_i) \frac{y_i^2}{\pi_i^2} + 2 \sum\sum_{\substack{i,j\in s \\ j>i}} \frac{\pi_{ij}-\pi_i\pi_j}{\pi_{ij}} \frac{y_i y_j}{\pi_i\pi_j} . \qquad (4.2.35)$$

The ratio estimator y''_{CS}, formed analogously to (4.2.25), has approximate mean square error (see Appendix B) given by

$$V(y''_{CS}) \doteq \sum_{I=1}^{N} \pi_I(1-\pi_I) \left(\frac{Y_I}{\pi_I} - \frac{Y}{n}\right)^2 + 2 \sum\sum_{\substack{I,J=1 \\ J>I}}^{N} (\pi_{IJ}-\pi_I\pi_J) \left(\frac{Y_I}{\pi_I} - \frac{Y}{n}\right)\left(\frac{Y_J}{\pi_J} - \frac{Y}{n}\right) + P_{0C}Y^2 , \qquad (4.2.36)$$

where P_{0C} is the probability of selecting an empty sample.

The conventional estimator of this approximate mean square error is

$$v(y''_{CS}) = \sum_{i\in s} (1-\pi_i) \left(\frac{y_i}{\pi_i} - \frac{y''}{n}\right)^2 + 2 \sum\sum_{\substack{i,j\in s \\ j>i}} \frac{\pi_{ij}-\pi_i\pi_j}{\pi_{ij}} \left(\frac{y_i}{\pi_i} - \frac{y''}{n}\right)\left(\frac{y_j}{\pi_j} - \frac{y''}{n}\right) + P_{0C}y''^2 . \qquad (4.2.37)$$

A more stable estimator than (4.2.37) may be obtained by multiplying the first term of (4.2.37) by n/m and the second term by $n(n-1)/m(m-1)$.

The expressions for π_{IJ} and for P_{0C} were devised by Brewer, Early and Hanif (1980) (see Appendices D and E). They are most conveniently expressed in terms of a population listed in ascending order of size, so that $\pi_1 \le \pi_2 \le \ldots \le \pi_N$. Writing $N\pi_I = [N\pi_I] + K_I$ where $[A]$ denotes the integral part of A , they showed that

$$\pi_{IJ} = \frac{[N\pi_I](N\pi_J-1)+K_I[N\pi_J]+\max\{(K_J-K_I),0\}}{N(N-1)} . \qquad (4.2.38)$$

Clearly also

$$P_{0C} = \int_{\theta=0}^{1} \{P_{0C}|\theta\} d\theta \ , \tag{4.2.39}$$

where $\{P_{0C}|\theta\}$ is the probability of an empty sample given a particular value of θ .

Approximate expressions for π_{IJ} and P_{0C} can be obtained on the assumption that the π_I are integer multiples of N^{-1} , in which case $k_I = 0$ for all I , *viz.*

$$\pi_{IJ} = \frac{\pi_I(N\pi_J-1)}{N-1} \ , \tag{4.2.40}$$

and

$$\pi_{IJ} - \pi_I\pi_J = \frac{\pi_I(\pi_J-1)}{N-1} \ , \tag{4.2.41}$$

where $\pi_J > \pi_I$.

Even when the π_I are not all integer multiples of N^{-1} , the use of (4.2.41) for $(\pi_{IJ}-\pi_I\pi_J)$ in the formulae (4.2.34) and (4.2.36) results in good approximations for the variance of y'_{CS} and the mean square error of y''_{CS} . With this same approximate formula for π_{IJ} , P_{0C} becomes

$$P_{0C} = \begin{cases} \dfrac{1}{N!} \displaystyle\prod_{I=1}^{N} (I-N\pi_I) & \text{if } \displaystyle\min_I (I-N\pi_I) > 0 \ , \\ 0 & \text{otherwise.} \end{cases} \tag{4.2.42}$$

This P_{0C} is much smaller than the corresponding P_0 for Poisson sampling.

4.2.7 Lahiri's Estimator

The use of Procedures 45 and 46, for which the probability of selection of a sample is proportional to its total measure of size, ensure that the conventional ratic estimator

$$y'' = \left\{\left(\sum_{i=1}^{n} y_i\right) \Big/ \left(\sum_{i=1}^{n} z_i\right)\right\} Z \tag{4.2.43}$$

is unbiased.

When n is large, the probabilities of selecting all particular samples tend to equality, and the variance of y'' may be evaluated and estimated using the Taylor expansion expressions familiar from equal probability sampling.

When, however, n is small and the inequalities in selection probability cannot be ignored, difficulties arise in the estimation of variance. For $n = 2$ the variance is

$$V(y'') = \frac{1}{N-1} \sum_{\substack{I,J=1 \\ J>I}}^{N}\sum \frac{(Y_I+Y_J)^2}{P_I+P_J} - Y^2 . \qquad (4.2.44)$$

Rao and Bayless (1969) used model (1.8.5) and obtained the following expression for the expected variance of (4.2.43) for $n = 2$:

$$E^*V(y'') = \sigma^2 Z^2 (N-1)^{-1} \sum_{J>I}^{N}\sum \frac{P_I^{2\gamma}+P_J^{2\gamma}}{P_I+P_J} - \sum_{I=1}^{N} P_I^{2\gamma} . \qquad (4.2.45)$$

They also found that for $n = 2$ the Lahiri estimator was more efficient than the Horvitz-Thompson, Raj, Murthy and RHC estimators when either

(a) few units in the population had large sizes relative to the sizes of remaining units in the population, and samples containing those units gave good estimates of Y , or

(b) the coefficient of variation of the benchmark variable was small.

For other populations it had poor efficiency. Raj (1954) and Sen (1955) provided an unbiased variance estimator for $n = 2$, namely

$$v_{\mathrm{R}}(y'') = y''^2 - \frac{1}{P_1+P_2}\left[(y_1-y_2)^2+2Ny_1y_2\right] . \qquad (4.2.46)$$

This can take negative values, and was found by Rao and Bayless (1969) to have very poor stability. The modification suggested by Sen (1955), replacing negative values of $v_{\mathrm{R}}(y'')$ by zero, did not lead to any substantial improvement.

Bayless and Rao (1970) in extending their earlier investigations to the cases $n = 3$ and $n = 4$, arrived at the same conclusions as for $n = 2$, both with respect to the efficiency of the estimator of total and the poor performance of the variance estimators. The reader is referred to that paper for variance formulae and estimators.

More recently Rao and Vijayan (1977) have proposed two new unbiased variance estimators which for some populations are nonnegative. For the case $n = 2$ these estimators coincide and take the form

$$v_a(y'') = -a_{12}(s)z_1z_2\left(\frac{y_1}{z_1} - \frac{y_2}{z_2}\right)^2 , \qquad (4.2.47)$$

where

$$a_{12}(s) = \frac{Z}{z_1+z_2}\left[\frac{Z}{z_1+z_2} - (N-1)\right] . \qquad (4.2.48)$$

For $n > 2$ the first estimator suggested is

$$v_{a1}(y'') = - \sum_{\substack{i,j \in s \\ j>i}}\sum \frac{a_{ij}}{\pi_{ij}} z_i z_j \left(\frac{y_i}{z_i} - \frac{y_j}{z_j}\right)^2 \tag{4.2.49}$$

where

$$\pi_{ij} = \frac{(n-1)(N-n)}{(N-1)(N-2)} \frac{z_i+z_j}{Z} + \frac{(n-1)(n-2)}{(N-1)(N-2)} \tag{4.2.50}$$

and

$$a_{ij} = Z \sqrt{\binom{N-1}{n-1}} \sum_{s \ni i,j} \sqrt{\left(\sum_{k \in s} z_k\right)} - 1 . \tag{4.2.51}$$

The second estimator suggested is

$$v_{a2}(y'') = Z\sqrt{\left(\sum_{k\in s} z_k\right)}\left(\frac{N-1}{n-1} - Z\sqrt{\left(\sum_{k\in s} z_k\right)}\right) \sum_{\substack{i,j\in s \\ j>i}} z_i z_j \left(\frac{y_i}{z_i} - \frac{y_j}{z_j}\right)^2 . \tag{4.2.52}$$

This second estimator is computationally simpler than the first, but is consistently less efficient, and usually has a greater probability of producing a negative estimate. Both estimators are typically (though not invariably) much more efficient than $v_R(y'')$.

4.3. COMPARISON OF SAMPLING SCHEMES USING SPECIAL ESTIMATORS

The criteria for comparison will be as in Chapter 3; limitation to the case $n > 2$, simplicity in selection, simplicity in variance estimation, the efficiency of the estimator of total, the unbiasedness and stability of the variance estimator and rotatability. All these concepts have been described in Chapter 3.

4.4. LIMITATION TO SAMPLE SIZE $n = 2$

When the Horvitz-Thompson estimator was used, the limit to the number of units which could be selected in sample was $P_{\max}^{-1} = Z/Z_{\max}$. This was because the probability of its inclusion in sample, $nP_{\max}$, was not allowed to exceed unity. For four of the seven sampling procedures considered in this Chapter (Procedures 4, 25, 45 and 46) this limit is not relevant and the number of units in sample can be set at any value up to N itself. For the Poisson sampling group (Procedures 27, 38 and 39) the upper limit to the expected number of sample units remains $P_{\max}^{-1}$.

4.5. SIMPLICITY IN SELECTION PROCEDURE

It was mentioned in Chapter 3 that systematic procedures have an obvious advantage over all other procedures in simplicity of selection procedure. The Raj and Murthy sample schemes use a selection procedure which is hardly more complicated than systematic selection. The RHC Procedure 25 involves the formation of n random groups. It is therefore slightly more tedious than that of the Raj and Murthy schemes but perhaps slightly easier to apply than the rejective Rao-Sampford Procedure 11.

Poisson Sampling uses a series of N binomial trials to determine whether each population unit is to be included in sample or not. Although this is more tedious than the procedures mentioned above, it is appreciably simpler to use than those which involve iteration. Collocated sampling is not feasible without a computer, at least for the comparatively large populations for which it was devised.

The Ikeda-Midzuno Procedure 46 appears to be somewhat less cumbersome than Lahiri's Procedure 45 if a sample selected with probability proportional to aggregate size is desired.

4.6. SIMPLICITY IN VARIANCE ESTIMATION

In Chapter 3 it was pointed out that for procedures using the Horvitz-Thompson estimator, the problem of estimating variance was virtually identical with the problem of determining the π_{IJ} . In consequence it was the case for most, though not all, procedures that the simplicity of the variance estimation procedure was directly related to the simplicity of the selection procedure.

For the estimators discussed in this Chapter the variance estimation formulae for these procedures have already been set out in equations (4.2.5), (4.2.10), (4.2.19), (4.2.24), (4.2.27), (4.2.35) and (4.2.37). It will be seen that Raj's estimator, the RHC estimator and the estimators for Poisson Sampling all have quite simple variance estimators for any n . The same is true for collocated sampling provided the approximate formula (4.2.40) is used. The Murthy variance estimator is simple for $n = 2$ but becomes rapidly more complicated as n increases.

It has already been mentioned that $v_{a2}(y'')$ is simpler than $v_{a1}(y'')$ for estimating the variance of Lahiri's estimator.

4.7. EFFICIENCY OF ESTIMATOR OF TOTAL

In this Section the efficiency of the various procedures will be considered empirically and semi-empirically using the model (1.8.5).

In order to facilitate comparison with the formulae relevant to the use of the Horvitz-Thompson estimator with exact selection procedures, the symbols π_I, π_J, and so on, will be used to denote nP_I, nP_J and so on. Note that there are *not* the probabilities of inclusion in the sample for either the Raj-Murthy or for the Rao-Hartley-Cochran Procedures although they do have this meaning in the Poisson sampling group.

4.7.1 The Raj and Murthy Estimators

Writing π_I for $2P_I$ in (4.2.3), (4.2.4), (4.2.8) and (4.2.9), we have for $n = 2$:

$$t_{\text{mean}} = \tfrac{1}{2}\left[(2+\pi_1)\frac{y_1}{\pi_1} + (2-\pi_1)\frac{y_2}{\pi_2}\right] , \tag{4.7.1}$$

$$V(t_{\text{mean}}) = \tfrac{1}{8} \sum_{\substack{I,J=1 \\ J\neq I}}^{N}\sum \pi_I\pi_J(n-\pi_I-\pi_J)\left(\frac{Y_I}{\pi_I} - \frac{Y_J}{\pi_J}\right)^2 , \tag{4.7.2}$$

$$t_{\text{symm}} = \frac{2}{4-\pi_1-\pi_2}\left[(2-\pi_2)\frac{y_1}{\pi_1} + (2-\pi_1)\frac{y_2}{\pi_2}\right] , \tag{4.7.3}$$

and

$$V(t_{\text{symm}}) = \tfrac{1}{2} \sum_{\substack{I,J=1 \\ J\neq I}}^{N}\sum \pi_I\pi_J \frac{2-\pi_I-\pi_J}{4-\pi_I-\pi_J}\left(\frac{Y_I}{\pi_I} - \frac{Y_J}{\pi_J}\right)^2 . \tag{4.7.4}$$

Rao and Bayless (1969) used the model (1.8.5) and obtained the expected variances of the estimators t_{mean} and t_{symm}, from (4.7.2) and (4.7.4), *viz.*

$$E^*[V(t_{\text{mean}})] = \frac{\sigma^2}{4}(Z/2)^{2\gamma} \sum_{\substack{I,J=1 \\ J\neq I}}^{N}\sum \pi_I^{2\gamma-1}\pi_J(4-\pi_I-\pi_J) \tag{4.7.5}$$

and

$$E^*[V(t_{\text{symm}})] = \sigma^2(Z/2)^{2\gamma} \sum_{\substack{I,J=1 \\ J\neq I}}^{N}\sum \pi_I^{2\gamma-1}\pi_J \frac{2-\pi_I-\pi_J}{4-\pi_I-\pi_J} . \tag{4.7.6}$$

Hanurav (1966b) and Vijayan (1966) compared the relative efficiencies under the model (1.8.5) of the Horvitz-Thompson estimator, y'_{HT} and t_{symm} . They proved that the Horvitz-Thompson estimator was more efficient than t_{symm} for $\gamma = 1$. Rao (1966b) further proved that y'_{HT} was more efficient than y'_{HH} for all values of γ . Rao (1966b) and Vijayan (1966) also proved that the t_{symm} was better than y'_{HT} for $\gamma = 0.5$.

Extensive empirical and semi-empirical studies were carried out by Rao and Bayless (1969) for the case $n = 2$, and by Bayless and Rao (1970) for the cases $n = 3$ and $n = 4$. In their empirical studies they found that Murthy's estimator was nearly always more efficient than the Horvitz-Thompson estimator, except in certain artificial populations. In their semi-empirical studies of the case $n = 2$, the values of γ which they used were 0.5, 0.75, 0.875 , and 1.0 . For all these values of γ , Murthy's procedure was consistently more efficient than Raj's procedure. Raj's estimator was usually more efficient than the Horvitz-Thompson estimator for $\gamma = 0.5$ and usually less efficient for $\gamma > 0.5$. Murthy's estimator was more efficient than the Horvitz-Thompson estimator for $\gamma \le 0.875$ and less efficient for $\gamma = 1.0$. For $\gamma = 0.875$ Murthy's estimator was nearly always the more efficient but the difference was very small.

For $n = 3$ and 4 Bayless and Rao investigated the cases $\gamma = 0.75$, 0.875 and 1.0 only. Raj's estimator was less efficient than the Horvitz-Thompson estimators in almost every case. Murthy's estimator was again more efficient than the Horvitz-Thompson estimator for $\gamma \le 0.875$.

Few of the differences in efficiency between Murthy's estimator and the Horvitz-Thompson estimator for natural populations exceeded 10% . The same was true for the comparison of the Raj and Horvitz-Thompson estimators.

The close agreements between the empirical and the semi-empirical results of Rao and Bayless tend to suggest that the form of the linear stochastic model assumed by them is reasonably appropriate. However, Samiuddin *et al* (1978) studied the behaviour of t_{symm}, y'_{HT} and several other estimators with six semi-empirical and six artificial populations. The Horvitz-Thompson estimator was found to be reasonably efficient in all cases. Murthy's estimator was reasonably efficient for the semi-empirical populations but somewhat less satisfactory for the artificial ones.

4.7.2 The Rao-Hartley-Cochran Estimator

When N is a multiple of n , the Rao-Hartley-Cochran variance estimator attains the minimum value (4.2.18). The expected variance of the RHC estimator is (Rao and Bayless, 1969)

$$E^*V\left(y'_{\text{RHC}}\right) = \sigma^2 c_0 c_1 (Z/2)^{2\gamma} \sum_{I=1}^{N} \left(2-\pi_I\right)\pi_I^{2\gamma-1} , \qquad (4.7.7)$$

where

$$c_0 = \frac{N_1^2+N_2^2-N}{N^2-N_1^2-N_2^2} \quad \text{and} \quad c_1 = \frac{N^2-N_1^2-N_2^2}{N(N-1)} .$$

A corresponding formula for $n > 2$ is given by Bayless and Rao (1970).

In Rao's (1966b) comparison of the relative efficiencies of various estimators under the same model, he showed that the RHC estimator was less efficient than both Murthy's estimator and the Horvitz-Thompson estimator for $\gamma = 1$. The Horvitz-Thompson estimator was more, equally or less efficient than the RHC estimator according as γ was greater than, equal to, or less than 0.5 respectively. Further comparisons of the efficiency of the RHC estimator with that of the Horvitz-Thompson estimator are given by Pedgaonkar and Prabhu Ajgaonkar (1978). Pathak (1966) also proved that for large N the RHC estimator is less efficient than Murthy's estimator for $\gamma \geq 0.5$. Singh and Kishore (1975) showed that after taking expected cost into account the Hansen-Hurwitz estimator based on multinomial sampling was sometimes superior to the RHC estimator.

Rao and Bayless (1969) and Bayless and Rao (1970) in their empirical studies for $n = 2$, 3 and 4 concluded that the RHC estimator was consistently less efficient than Murthy's estimator, and that it was sometimes slightly more and sometimes slightly less efficient than the Horvitz-Thompson estimators.

In the semi-empirical studies carried out by the same authors, the RHC estimator was found to be consistently less efficient than both the Murthy and the Horvitz-Thompson estimators. Its efficiency vis-à-vis the Murthy estimator was not greatly affected by the value of γ, but vis-à-vis the Horvitz-Thompson estimators it was least efficient for $\gamma = 1$. As with Murthy's and Raj's estimators, most of the differences were only of the order of a few percent, except for $n = 4$ where differences of 20% and 30% were not uncommon.

4.7.3. Poisson and collocated sampling

Two empirical populations were used by Brewer, Early and Hanif (1980) to compare Poisson and collocated sampling with other unequal probability sampling strategies. The first of these was the population of 49 cities listed in Cochran (1963), p. 156, and the second that of 270 blocks listed in Kish (1965), p. 624. The Cochran population contained one exceptional unit with very low π_I and high ratio Y_I/π_I. The Kish population contained no such maverick.

The strategies compared were as follows:

(i) Sampling with replacement (that is multinomial sampling) with the Hansen-Hurwitz (1943) estimator.

(ii) Sampling without replacement (m fixed) with the Horvitz-Thompson (1952) estimator. For this strategy the asymptotic variance formula

$$V(y'_{\text{HT}}) \doteqdot \sum_{I=1}^{N} \pi_I\left(1 - \frac{n-1}{n}\pi_I\right)\left(\frac{Y_I}{\pi_I} - \frac{Y}{n}\right)^2 \tag{1.8.4}$$

was used.

(iii) Poisson sampling with the unbiased estimator y'_{PS} .

(iv) Poisson sampling with the ratio estimator y''_{PS} .

(v) Collocated sampling with the unbiased estimator y'_{CS} .

(vi) Collocated sampling with the ratio estimator y''_{CS} .

For Poisson and collocated sampling, variances were calculated both excluding and including the terms P_0Y^2, $P_{0C}Y^2$, so as to indicate the importance of the non-zero probability of an empty sample.

For collocated sampling the mean square errors were calculated using

(a) the exact π_{IJ} values given in (4.2.38),

(b) the approximate π_{IJ} values given by (4.2.40),

(c) the approximate π_{IJ} values given by (4.2.40) wherever these exceeded zero, but otherwise replaced by zero.

In every case the probabilities of inclusion in sample were taken to be proportional to the Z-values supplied. The use of the approximate formula (4.2.40) for the π_{IJ} resulted in reasonable approximations for the variance and mean square error formulae for collocated sampling. The better of the two approximations was achieved when the negative values obtained from (4.2.40) are set equal to zero, but the advantage held only when n is small.

The results based on exact π_{IJ} values are given in Tables 4.1 and 4.2. Some highlights of these are as follows:

1. When the probability of an empty sample is small or zero, the mean square error of the ratio estimator for Poisson or collocated sampling is comparable with the variance of the Horvitz-Thompson estimator when m is fixed. (The calculations actually show the ratio estimator mean square error to be smaller, but this is due to the Taylor series approximation.)

2. When the probability of an empty sample is of the order of 0.003 or greater, the contribution to the variance from the empty sample term is too large to be ignored.

3. The probability of an empty sample is at least an order of magnitude smaller in collocated sampling than in Poisson sampling, and becomes exactly zero for large samples.

TABLE 4.1

Comparisons of Efficiencies for Different Strategies with Hansen-Hurwitz *ppswr* as Standard

Strategy	Cochran's Population $N = 49$			Kish's Population $N = 270$			
	$n = 2$	5	9	$n = 2$	10	20	30
Hansen-Hurwitz *ppswr*	1.0000	1.0000	1.0000	1.0000	1.0000	1.0000	1.0000
Horvitz-Thompson π*pswor* (fixed sample size)	1.0037	1.0151	1.0307	1.0054	1.0511	1.1144	1.1859
Poisson							
- Unbiased	0.1812	0.2022	0.2391	0.0817	0.0877	0.0967	0.1078
- Ratio							
- Ignoring P_0Y^2	1.0075	1.0109	1.0346	1.0109	1.0571	1.1212	1.1935
- Including P_0Y^2	0.4534	0.9380	1.0344	0.2472	1.0532	1.1212	1.1935
Collocated							
- Unbiased	0.3127	0.3627	0.4300	0.1303	0.1440	0.1592	0.1776
- Ratio							
- Ignoring $P_{0C}Y^2$	1.0169	1.0295	1.0472	1.0744	1.1519	1.2375	1.3287
- Including $P_{0C}Y^2$	0.8745	1.0292	1.0472	0.5771	1.1519	1.2375	1.3287

TABLE 4.2

Comparisons of Probabilities of Empty Samples for Poisson and Collocated Sampling

Strategy	Cochran's Population, $N = 49$			Kish's Population, $N = 270$		
	$n = 2$	5	9	$n = 2$	10	20
Poisson, P_0	0.1237	0.3459×10^{-2}	0.4139×10^{-5}	0.1333	0.3055×10^{-4}	0.3666×10^{-9}
Collocated, P_{0C}	0.1633×10^{-1}	0.1155×10^{-4}	zero	0.3499×10^{-1}	0.2787×10^{-7}	0.9240×10^{-16}

NOTE: The probabilities of empty samples for the above tables have been calculated using the approximation (E.2) from Appendix E.

4.7.4. Lahiri's Estimator

Except when n is very small, sampling with probability proportional to aggregate size approximates sampling with equal probabilities, Cochran (1953) showed that the conventional ratio estimator with equal probability sampling was more efficient than the Horvitz-Thompson estimator with $\pi pswor$ for low values of γ with a break-even point close to $\gamma = \frac{1}{2}$. While much the same kind of conclusion was reached for the RHC estimator, the contrast here is much more severe. The RHC estimator closely resembles the Horvitz-Thompson estimator, while the conventional ratio estimator is entirely different. Similarly the RHC Procedure 25 is nearly an exact $\pi pswor$ scheme, while Procedures 45 and 46 approximate equal probability sampling.

4.8. UNBIASEDNESS AND STABILITY OF VARIANCE ESTIMATORS

4.8.1. The Raj and Murthy Estimators

Rao and Bayless (1969) used the model (1.8.5) to find the stability of variance estimators (4.2.5) and (4.2.10). They had shown that the leading terms in the expected variances of (4.2.5) and (4.2.10) for $n = 2$ were

$$E^*\left[Ev^2\left(t_{\text{mean}}\right)\right] = \frac{3}{64}\sigma^4(Z/2)^{4\gamma} \sum_{\substack{I,J=1 \\ J\neq I}}^{N}\sum \pi_I\pi_J\left(2-\pi_I\right)^3\left(\pi_I^{2\gamma-2}+\pi_I^{2\gamma-2}\right)^2 , \tag{4.8.1}$$

and

$$E^*\left[Ev^2\left(t_{\text{symm}}\right)\right] = 3\sigma^4(Z/2)^{4\gamma} \sum_{\substack{I,J=1 \\ J\neq I}}^{N}\sum \frac{\pi_I\pi_J\left(2-\pi_I\right)\left(2-\pi_J\right)\left(2-\pi_I-\pi_J\right)}{\left(4-\pi_I-\pi_J\right)^3}\left(\pi_I^{2\gamma-2}+\pi_J^{2\gamma-2}\right)^2 . \tag{4.8.2}$$

The leading term in the expected variance of (4.2.11) is also presented here:

$$E^*\left[Ev_M^2\left(t_{\text{symm}}\right)\right] = \frac{3}{128}\sigma^4(Z/2)^{4\gamma} \sum_{\substack{I,J=1 \\ J\ I}}^{N}\sum \pi_I\pi_J\left(4-\pi_I-\pi_J\right)\left(2-\pi_I\right)\left(2-\pi_J\right)\left(\pi_I^{2\gamma-2}+\pi_J^{2\gamma-2}\right)^2 . \tag{4.8.3}$$

Rao and Bayless (1969) and Bayless and Rao (1370) made semi-empirical and empirical studies of the stabilities of variance estimators for $n = 2, 3$, and 4. They concluded from their semi-empirical studies that Murthy's variance estimator was consistently more stable than the Sen-Yates-Grundy variance estimator. This was particularly the case for the smaller values of γ. Murthy's variance estimator also tended to be more stable than Raj's variance estimator, especially for the larger values of γ and of n.

In their empirical studies Rao and Bayless concluded that Raj's and Murthy's variance estimators were essentially equivalent in stability for $n = 2$, but that

Murthy's was usually slightly more stable for $n = 4$. Both these variance estimators were almost always more stable than the Sen-Yates-Grundy variance estimator, and the gains were often appreciable.

4.8.2. The Rao-Hartley-Cochran Estimator

Rao and Bayless (1969) and Bayless and Rao (1970) used the linear stochastic model (1.8.5) to derive the expected variance of the RHC variance estimator. The formulae, which are extremely complicated, are given in their 1969 paper for $n = 2$ and in Appendix B of their 1970 paper for any n .

In their semi-empirical studies they also concluded that for $n = 2$, the RHC variance estimator was consistently more stable than the Raj, Murthy, and Sen-Yates-Grundy estimators for all values of γ ; however the gains over Murthy's variance estimator were not large. For $n = 3$ and 4 , the RHC variance estimator was still almost always more stable than the Murthy variance estimator for $\gamma = 0.875$, but for $\gamma = 1$ the reverse was the case. It was consistently more stable than the Sen-Yates-Grundy variance estimator for all values of γ . In their empirical studies they found that for all values of n considered, the RHC variance estimator was more stable than the Raj, Murthy, and Sen-Yates-Grundy variance estimators.

These special variance estimators are much more stable than the Sen-Yates-Grundy variance estimator, even when the joint probabilities of selection are chosen specifically to stabilise the latter. This result is consonant with Raj's own findings (1956a) and is also heuristically plausible in that the coefficients of $\left((y_1/\pi_1)-(y_2/\pi_2)\right)^2$ for all these three variance estimators are usually close to and always less than unity, whereas for the Sen-Yates-Grundy variance estimator the coefficients are $\left(\pi_1\pi_2\pi_{12}^{-1}-1\right)$, which tend to be rather variable (see for instance Table 3.1).

4.8.3. Lahiri's Estimator

It has already been mentioned that $v_{a1}(y'')$ is appreciably more stable than $v_{a2}(y'')$ for estimating the variance of Lahiri's estimator.

4.9. ROTATABILITY

Raj's and Murthy's sample schemes are not appropriate for rotation except using the Alternative III mentioned in Section 3.8. Samples selected by the Rao, Hartley and Cochran Procedure 25 may be rotated using a slightly modified version of Alternative I. Since selection within each of the n groups occurs independently, each selection may be rotated around the population units allocated to that group, starting from a random point within the first unit selected (for the reason indicated in Section 3.8).

A method of Poisson sampling which allows for rotation and updating in a simple way was presented by Brewer, Early and Joyce (1972).

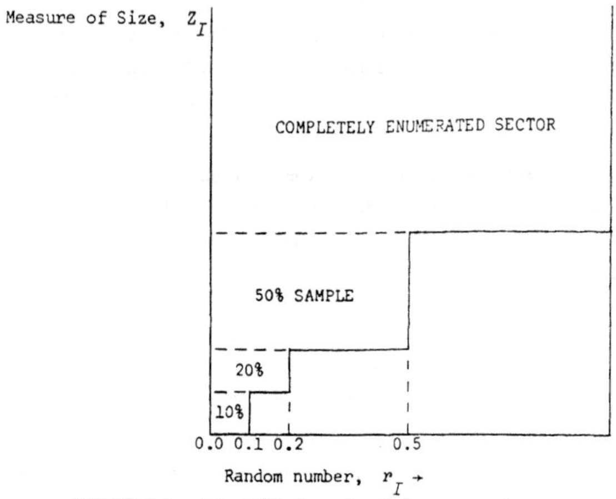

FIGURE 4.1. Stratified random Poisson sample.

Figure 4.1 shows how Poisson sampling works for a stratified random sample with three sampled strata and a completely enumerated sector. The units of the population correspond to points on the chart specified by r and Z. The sample consists of all points in the chart to the left of the thick line. Rotation can be effected by shifting the sample area to the right. If the shift is 0.02 in r, all units for which r is less than 0.02 are rotated out of sample and replaced by units for which r lies between $\pi(Z_I)$ and $\pi(Z_I) + 0.02$ where $\pi(Z_I)$ is the probability of inclusion in sample of a unit with size Z_I. This would give a 20% rotation in the lowest sampled stratum, 10% in the next, and 4% in the highest. The chart should be thought of as cylindrical, so that for the completely enumerated sector where $\pi(Z_I) = 1.00$, the new limit of $\pi(Z_I) + 0.02$ or 1.02 brings in again those units which would otherwise be rotated out, giving nil rotation in the completely enumerated sector.

Figure 4.1 can obviously be used to select other samples of various sizes with minimum or maximum overlap, and shows at a glance what is feasible and what is not feasible about, say, different rates of rotation for samples with minimum overlap.

Figure 4.2 illustrates two different ways of rotating a sample drawn with probabilities proportional to size (πps). The original sample is that of all points

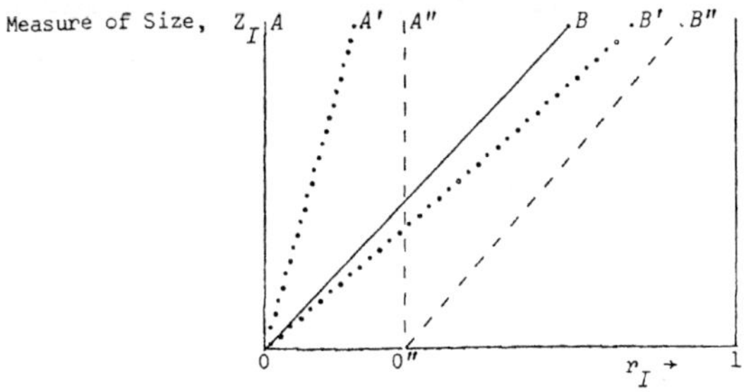

FIGURE 4.2. Two ways of rotating a πps Poisson sample.

contained in the area AOB . A fixed proportion rotation, such as 20% , gives the new sample area as $A'OB'$. A usually preferable alternative is represented by $A''O''B''$. This gives fast rotation for small units and slow for large. Similar procedures may be used if the probability of selection is any function of size.

A formal description of this method was presented by Brewer, Early and Hanif (1980). Choose an arbitrary fixed number c and (for all I) a uniformly distributed random number r_I in the interval $[0, 1)$. Then the Ith unit is in sample if $r_I < \max\{0, \pi_I - 1 + c\}$ or $c \le r_I < \min\{\pi_I + c, 1\}$; that is, the Ith unit is selected if $\{\pi_I, r_I\}$ lies in the shaded area in Figure 4.3. Since r_I is uniformly distributed over $[0, 1)$, the probability that the Ith unit is selected is $\max\{0, \pi_I - 1 + c\} + \min\{\pi_I + c, 1\} - c = \pi_I$ for all I , as required.

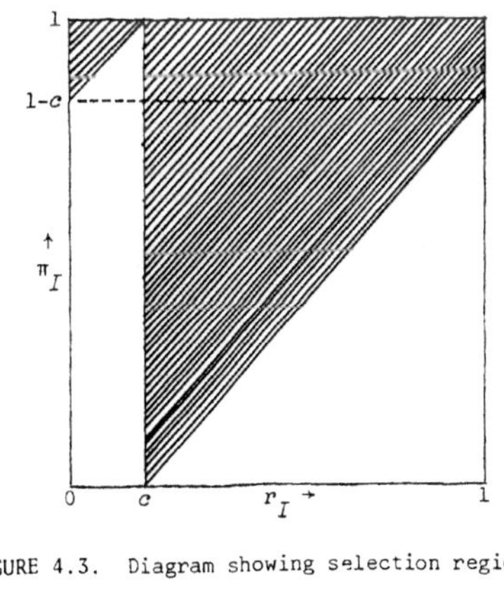

FIGURE 4.3. Diagram showing selection region.

In selecting a later sample with (updated) values π'_I for each I, we can control the expected overlap between the samples by varying the value of c. Suppose the value of c is altered for the revised sample to $c + d$ where $0 \le c+d < 1$ and $d > 0$. Then (see Appendix C) the probability of inclusion of the Ith unit in both samples, $B_I(d)$, takes the values set out in the following table.

	$d \ge \pi_I$	$d \le \pi_I$
$d \le 1-\pi'_I$	0	$\min\{\pi'_I, \pi_I-d\}$
$d \ge 1-\pi'_I$	$\min\{\pi_I, \pi'_I-1+d\}$	$\pi'_I+\pi_I-1$

(4.9.1)

Noticing that when $d \le \pi_I$ and $d \ge 1-\pi'_I$,

$$\pi'_I + \pi_I - 1 = \min\left(\pi_I, \pi'_I-1+d\right) + \min\left(\pi'_I, \pi_I-d\right),$$

we have that

$$\sum_{I=1}^{N} B_I(d) = \sum_{\pi_I \ge d} \min\left(\pi'_I, \pi_I-d\right) + \sum_{\pi'_I \ge 1-d} \min\left(\pi_I, \pi'_I-1+d\right) \qquad (4.9.2)$$

is the expected number of population units in the overlap of the samples, so that $n^{-1} \sum_I B_I(d)$ will be a measure of the expected proportion of original sample units retained. The maximum expected overlap is $\sum_I \min\left(\pi_I, \pi'_I\right)$ which occurs if $d = 0, 1$. The minimum expected overlap will be approximated closely in practice by taking $d = \frac{1}{2}$, and its value will be close to $\sum_{\pi_I+\pi'_I>1} \left(\pi_I+\pi'_I-1\right)$ (see Appendix C).

These maximum and minimum overlap situations may be represented diagramatically as follows:

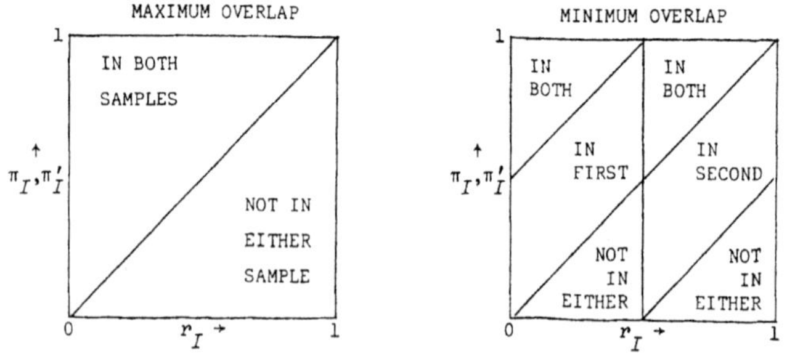

FIGURE 4.4. Diagrams showing maximum and minimum overlap.

An alternative method of achieving minimum overlap (when the π_I and π_I' are comparable) is illustrated in Figure 4.5.

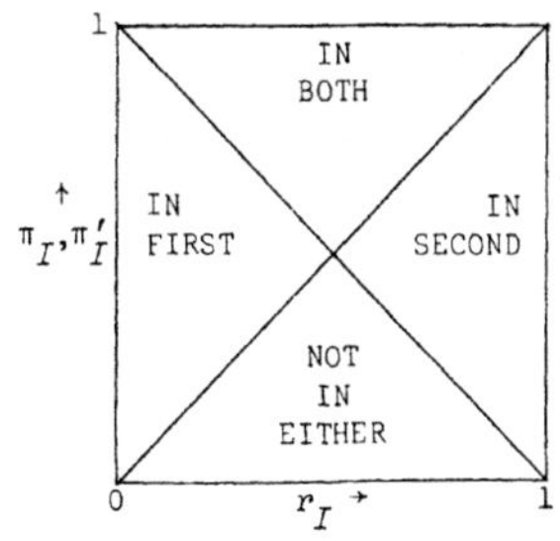

FIGURE 4.5. An alternative method of achieving minimum overlap.

This second method of achieving minimum overlap is more powerful than the first in situations where the π_I and π_I' are roughly proportional but differ substantially in magnitude. To simplify the discussion suppose that $\pi_I' = \frac{1}{2}\pi_I$ for all I for which π_I is definable, and that π_I' goes on increasing to a maximum of unity for large units which in the first survey lie in the completely enumerated (CE) sector. In this situation maximum and minimum overlaps are achieved by using the diagrams set out in Figure 4.6. The annotations in these diagrams indicate whether the regions are CE, 'IN' sample or 'OUT' of sample on each occasion.

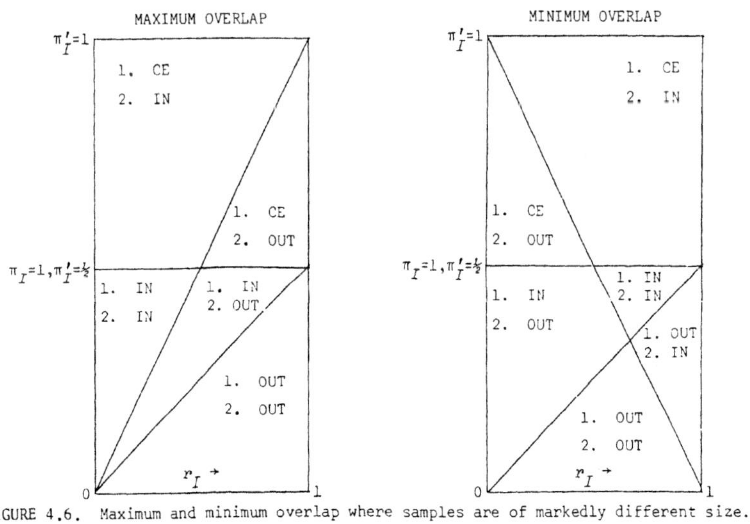

FIGURE 4.6. Maximum and minimum overlap where samples are of markedly different size.

The flexibility which can be achieved in the control of rotation and respondent burden using such diagrams is obvious. They are applicable only to Poisson and collocated sampling. It will be suggested in Chapter 6 that these procedures are for this reason particularly appropriate for use in large scale surveys of businesses, farms, factories, and so on, where units differ greatly in size and it is not unusual to have several samples in use simultaneously which have been selected from the same population.

Another approach to achieving a desired overlap between successive Poisson samples is to stipulate a new inclusion probability for each population unit, conditional on whether or not that unit was included in the first sample. This Keyfitz (1951) inspired method was used by the US Bureau of the Census (Ogus and Clark, 1971) and yields the same overlap as the method first described, but does not share its simple control properties.

As already mentioned, the method of rotation and updating for controlling overlap between successive samples applies equally to collocated sampling. The only complication arises when units enter and leave the population between the selection of the original and revised samples. In these cases we must allow the values of r_I to change between samples. We can minimize this problem (improving on the procedure suggested by Brewer, Early, and Joyce, 1972) by defining a new ordering H to replace L as follows.

Assume that l new units have been added to the population and k original units deleted. There are two cases to consider.

Case 1: $l > k$

The population labels of the k deleted units are attached at random to k of the l new units. The remaining $l - k$ new units are assigned population labels $L_I = N+1, \ldots, N+l-k$. Choose, using simple random sampling without replacement, $l - k$ distinct integers $H_{N+1}, \ldots, H_{N+l-k}$ from the set $\{1, \ldots, N+l-k\}$. Then let the Jth largest of the remaining $N - k$ integers define the new label value H_I where I is the original population label of the surviving units with the Jth largest value of L_I .

Case 2: $l \leq k$

The population labels L_I of a random sample of l of the k deleted units are attached at random to the l new units. The remaining unattached $k - l$ population labels I are destroyed. Then let the integer J define the new label value H_I where I is the original population label of a new or surviving unit with the Jth largest value of L_I .

We have thus defined H_I for each unit in the new population, and the values of H are uniformly distributed over the set $\{1, \ldots, N+l-k\}$. For each I as above, set

$$t_I = \frac{\theta+H_I-1}{N+l-k} . \tag{4.9.3}$$

Provided k and l are small in comparison to N we may assume $t_I \doteq r_I$ for any surviving unit, and thus use the result of Poisson sampling to control overlap.

Collocated sampling thus retains the most desirable properties of Poisson sampling (simplicity of selection and estimation of variance, and control over the sample). The π_{IJ} can be evaluated quite straightforwardly and are given in (4.2.40).

4.10. SUMMARY

In Tables 4.3 and 4.4, summaries of the properties of the procedures using special estimators are given for $n = 2$ and $n > 2$ respectively.

TABLE 4.3

Summary of Properties of the Procedures using Special Estimators, $n = 2$
(with Rao-Sampford Procedure for Comparison)

Procedure	Raj	Murthy	RHC	Lahiri	Rao-Sampford
Is number in sample fixed?	yes	yes	yes	yes	yes
Is estimator unbiased?	yes	yes	yes	yes	yes
Efficiency* $\gamma < \frac{1}{2}$	close to standard	better than standard	better than standard	much better than standard	standard
Efficiency* $\frac{1}{2} < \gamma < 1$	below standard	about standard	below standard	much below standard	standard
Efficiency* $\gamma = 1$	much below standard	below standard	much below standard	much below standard	standard
Is variance estimator unbiased?	yes	yes	yes	yes	yes
Stability of variance estimator*	good	very good	excellent	unknown	standard
Simplicity in selection	excellent	excellent	very good	excellent	good
Simplicity in variance estimation	excellent	excellent	excellent	excellent	good
Which alternatives may be used for rotation?	III	III	II[1], III	III	II[2], III

* The efficiency of the Horvitz-Thompson estimator and of the Sen-Yates-Grundy variance estimator have been taken as the standard in these parts of the table.

[1] If Alternative II is used, the procedure is to replace the selected unit within each group in turn by a new unit selected PPS within the same group.

[2] If Alternative II is used, oversampling is needed.

TABLE 4.4

Summary of Properties of the Procedures using Special Estimators, $n > 2$
(with Rao-Sampford Procedure for Comparison)

Procedure	Raj	Murthy	PHC	Poisson (Unbiased Estimator)	Poisson (Ratio Estimator)	Collocated (Unbiased)	Collocated (Ratio)	Lahiri	Rao-Sampford
Is number in sample fixed?	yes	yes	yes	no	no	no	no	yes	yes
Limit on (expected) number in sample	N	N	N	Z/Z_{max}	Z/Z_{max}	Z/Z_{max}	Z/Z_{max}	N	Z/Z_{max}
Is estimator unbiased?	yes	yes	yes	yes	nearly	yes	nearly	yes	yes
Efficiency* $\gamma < \frac{1}{2}$	close to standard	better than standard	better than standard	much below standard	close to standard	much below standard	close to standard	much better than standard	standard
Efficiency* $\frac{1}{2} < \gamma < 1$	below standard	about standard	below standard	much below standard	close to standard	much below standard	close to standard	much below standard	standard
Efficiency* $\gamma = 1$	much below standard	below standard	much below standard	much below standard	close to standard	much below standard	close to standard	much below standard	standard
Is variance estimator unbiased?	yes	yes	yes	yes	nearly	yes	nearly	yes	yes
Stability of variance estimator*	good	very good	excellent	unknown	unknown	unknown	unknown	unknown	standard
Simplicity in selection	excellent	excellent	very good	fair	fair	needs computer	needs computer	excellent	good
Simplicity in variance estimation	excellent	poor	excellent	excellent	good	good[4]	good[4]	excellent	good
Which alternatives may be used for rotation?	III	III	II[1], III	I[2]	I[2]	I[2]	I[2]	III	II[3]

* The efficiency of the Horvitz-Thompson Estimator and of the Sen-Yates-Grundy variance estimator have been taken as the standard "average" values in these parts of the table.

[1] If Alternative II is used, the procedure is to replace the selected unit with each group in turn by a new unit selected PPS within the same group with replacement.

[2] If Alternative I us used, rotation is best specified using diagrams such as Figures 4.1 and 4.2.

[3] If Alternative II is used, over sampling is needed.

[4] Using approximate expressions for joint inclusion probabilities.

CHAPTER 5

MULTISTAGE SAMPLING

5.1. INTRODUCTION

Multistage sampling occupies a central role both in the theory and in the applications of unequal probability sampling. It was in the context of multistage sampling that unequal probability sampling was first suggested (Hansen and Hurwitz, 1943). There are in fact two quite different contexts in which unequal probability sampling is used:

1. Unistage unequal probability sampling of units such as factories, farms and businesses in large scale economic surveys. Here unequal probability sampling is used in place of stratification by size, and can be viewed as taking the size stratification idea to its logical conclusion (see Chapter 6).

2. Multistage unequal probability sampling in area surveys of individuals and households. Here multistage sampling is used partly to overcome the problem that lists of the ultimate sampling units are typically not available, and partly to reduce travel costs by ensuring that the sample units are geographically clustered. Unequal probability sampling, in this context, is used partly to reduce sampling errors and partly to ensure that the resulting area samples are conveniently structured.

The following example will serve to illustrate the point about having a convenient sample structure. Let us suppose that within a certain geographical structure there are eight Local Government Areas (LGAS) of different sizes, as measured by the most recent census population figures. (Column (1) of Table 5.1).

TABLE 5.1

Hypothetical structure of LGAS in geographical structure and selections based on Procedure 1.

LGA No.	Census Population	No. of Clusters (1) ÷ 50	Cumulated Clusters Cum(2)	Selection Points
	(1)	(2)	(3)	(4)
1	11000	220	220	
2	6000	120	340	
3	18000	360	700	622
4	17000	340	1040	
5	5000	100	1140	
6	7000	140	1280	
7	12000	240	1520	1422
8	4000	80	1600	
	80000	1600		

The design calls for first stage sample of two LGAS and a second stage sample of 100 sample households from the two LGAS combined. Considerations of design efficiency and simplicity in the processing require that the sample be self-weighting, that is, that every household in sample be given equal weight in the analysis.

An unbiased estimator of total having this self-weighting property may be achieved in many ways. The simplest way is to use equal selection probabilities at both stages. The sampling fraction at the first stage is $2/8 = 1/4$. The over all sampling fraction is $100/80{,}000 = 1/800$. Consequently the second stage sampling fraction must be $1/800 \div 1/4 = 1/200$. Unfortunately this results in very different second stage sample numbers depending on the result of the first stage selections. At one extreme, if the two largest LGAS, Nos 3 and 4, are selected, the second stage sample will consist of $35{,}000/200 = 175$ households. At the other extreme if the smallest pair, Nos 5 and 8, are selected, the number of households in the second stage sample will be only $9000/200 = 45$. This variability in sample size means that the unbiased estimator, despite its self-weighting property, is exceedingly inefficient. But any departure from this estimator results in the loss of either the unbiasedness or the self-weighting property. In fact no modified form of estimator can compensate for the selection of such a catastrophically inappropriate sample.

A much more satisfactory method of achieving a self-weighting unbiased estimator is to select with *πpswor* at the first stage, using an exact method, and then to choose a second stage sample within the selected LGAS in such a fashion that every household in the stratum has the same unconditional probability of inclusion in sample. Such a first stage selection has been undertaken in Table 5.1 using the ordered Systematic Procedure 1 for simplicity. The design calls for 100 sample households in all. Since administrative convenience typically demands a separate interviewer in each sample LGA, it is best if these sample households are distributed evenly between them. Consequently Column (2) in Table 5.1 divides the households in each LGA into potential sample clusters containing 50 households each. (In practice there would be some rounding here, and the cluster would not contain precisely 50 households.) Following Procedure 1, the number of clusters is cumulated in column (3) and a random number chosen between 1 and $1600/2 = 800$. In this case the random number chosen is 622. The sample selections are therefore made at the clusters numbered 622 and $622 + 800 = 1422$. This results in the selection LGAS Nos 3 and 7 at the first stage, and notionally of a cluster of 50 (in practice approximately 50) households within each of them.

Because the Horvitz-Thompson estimator is being used at the first stage and the weights at both stages are inversely proportional to the probabilities of inclusion in sample, the second stage sample fraction is simply the reciprocal of the number of clusters in each selected LGA. One of 360 clusters is to be selected in LGA No. 3, and hence the second stage sampling fraction in that LGA is $1/360$. For LGA No. 7 it is $1/240$. These fractions can be applied uniformly over the whole of each selected LGA, or one or more additional stages may be interposed between the selection of LGAS and the selection of households. Typical choices for such intermediate stages of selection are Census Collectors' Districts (Census Tracts) and street blocks. The selection procedure used at each such intermediate stage would then also be an exact *πpswor* procedure. If Procedure 1 or 2 is used for this purpose, selection follows exactly the same pattern as that given in Table 5.1.

While it is simplest to use the Horvitz-Thompson estimator (at all stages except the last) to achieve an unbiased self-weighting estimator for households, it is also possible to use the Raj or Murthy estimator with Procedure 4 or the RHC estimator with Procedure 25. The complication is that some departures from equal sized clusters are necessary in order to retain the self weighting property.

Suppose, to use the example of Table 5.1, that Procedure 4 had been used and had resulted in the selection of LGA No. 3 at the first draw and LGA No. 8 at the second. Whereas with an exact procedure and the Horvitz-Thompson estimator the second-stage sampling fractions could be read off immediately as $1/360$ and $1/80$ respectively, now for the Raj estimator they are

$$\frac{1}{360}\left(1+(18/80)\right) \doteq \frac{1}{293.9} \quad \text{for LGA No. 3}$$

and

$$\frac{1}{80}\left(1-(18/80)\right) \doteq \frac{1}{103.2} \quad \text{for LGA No. 8.}$$

The expected numbers of sample households are then 61.25 and 38.75 respectively. For the Murthy estimator the second stage sampling fractions are

$$\frac{1}{360}\,\frac{2\left(1-(4/80)\right)}{\left(2-(18/80)-(4/80)\right)} = \frac{1}{326.8} \quad \text{for LGA No. 3}$$

and

$$\frac{1}{80}\,\frac{2\left(1-(18/80)\right)}{\left(2-(18/80)-(4/80)\right)} = \frac{1}{89.0} \quad \text{for LGA No. 8.}$$

The expected numbers of sample housholds are then 55.1 and 44.9 respectively. This selection of the largest and the smallest units is the most extreme case. For other selections the departures would be smaller than these. (The fact that the reciprocals of the sampling fractions are no longer integers is an additional but minor inconvenience.)

More severe departures can be found with Procedure 25 and the RHC estimator. The most extreme random grouping is that of LGAS Nos 1, 3, 4 and 7 in the first group (total population 58,000) and LGAS Nos 2, 5, 6 and 8 in the second (total population 22,000). For self-weighting the clusters in the first group would then have to be of size 72.5 households and those in the second of size 27.5 households. Remembering that there is typically one interviewer in each selected LGA, this allocation could represent a considerable administrative problem.

It is for these reasons that the authors' personal preference is for the Horvitz-Thompson estimator. Its chief disadvantage, as pointed out in Chapters 3 and 4, is the instability of the Sen-Yates-Grundy variance estimator, even when the π_{IJ} are chosen to minimize its variance. The authors believe however that for most practical purposes the partly model-based variance estimator (3.7.4) can be used without serious error. (It would, however, be necessary to avoid the use of the ordered procedures, such as Procedure 1, in such circumstances. The simplest procedure for the present pusposes is unquestionably the Randomized Systematic Procedure 2.)

The situation in which the Raj, Murthy and RHC estimators are to be preferred is the one where high priority is given to accurate and purely design based variance estimation. Of these, the Murthy estimator requires the smallest departures from target sample allocations, but it is complicated to use except when $n = 2$. For $n > 2$ the effective choice is between the Raj estimator, which is relatively inefficient, and the RHC estimator, which can result in large departures from target sample allocations.

Finally, since there is no ideal way of proceeding using *upswor*, some may still prefer to use multinomial sampling and the Hansen-Hurwitz estimator. The variance reduction represented by the finite population correction is then entirely lost, but the simplicity of the selection and estimation procedures, and the further simplicity and stability of the variance estimation procedures leave little to be desired. In the next Section we shall find an additional reason for preferring the Hansen-Hurwitz estimator in the context of multistage sampling.

5.2. VARIANCE ESTIMATION FOR MULTISTAGE SAMPLING

A fundamental principle of multistage sampling is that when selection and estimation take place independently at the various stages, the variances of an unbiased estimator which arise from each of these different stages can be added. In particular the total variance of such a multistage estimator is equal to the variance arising from the first stage plus that arising from subsequent stages. Formally this may be written

$$V(y') = V_1(E_2 y') + E_1 V_2(y') \tag{5.2.1}$$

where E_1 denotes the expectation and V_1 the variance over all first stage samples, E_2 denotes the expectation over all second and subsequent stage samples, and $V_2(y')$ is the conditional variance of y' subject to the selection of a particular first stage sample. With an obvious extension of this notation, expression (5.2.1) for three stages may be written as

$$V(y') = V_1(E_2 y') + E_1 V_2(E_3 y') + E_1 E_2 V_3(y') \tag{5.2.2}$$

and for k stage sampling as follows:

$$V(y') = V_1(E_2 y') + E_1 V_2(E_3 y') + E_1 E_2 V_3(E_4 y') + \ldots + E_1 E_2 \ldots E_{k-1} V_k(y') \;. \tag{5.2.3}$$

If the totals for the first stage sample units were known exactly it would be possible to estimate $V_1(E_2 y')$, the first stage variance, in exactly the same fashion as in single stage sampling, and the same is true *mutatis mutandis* for the other stages. The basic problem of multistage variance estimation is that these totals have to be replaced by estimates from the lower stages of sampling, and that this introduces a component from these lower stages which in general bears no direct relationship to the actual variance from these lower stages.

5.2.1 Multinomial Sampling and the Hansen-Hurwitz Estimator

There is one exception to the statement in the previous paragraph. This exception occurs when the first stage sample is selected multinomially, the Hansen-Hurwitz estimator is used and a separate independent second stage sample of the same size is selected every time a given first stage unit appears in sample.

In this case the component from the lower stages of variance is exactly equal to the actual variance from these lower stages, and the estimator of variance which is obtained by replacing the totals of the first stage sample units by unbiased estimators of these totals is an unbiased estimator of the total variance. We may write the Hansen-Hurwitz estimator as

$$y'_{\text{HH}} = \frac{1}{n}\sum \frac{y'_i}{p_i} , \tag{5.2.4}$$

where y_i is the second and lower stage sample estimate of y_i based on the sample selected corresponding to the ith first stage sample draw. The variance of this estimator is

$$V\left(y'_{\text{HH}}\right) = \frac{1}{n}\sum_{I=1}^{N} P_I\left(\frac{Y_I}{P_I} - Y\right)^2 + \frac{1}{n}\sum_{I=1}^{N} \frac{V_2\left(Y'_I\right)}{P_I} , \tag{5.2.5}$$

where $V_2\left(Y'_I\right)$ is the second and lower stage variance of an unbiased estimator of Y_I based on the sample selected corresponding to a single first stage sample draw. But the variance estimator formed by replacing y_i by y'_i in (1.3.5) has expectation

$$E\,\frac{1}{n(n-1)}\sum_{i=1}^{n}\left(\frac{y'_i}{p_i} - y'_{\text{HH}}\right)^2$$

$$= \frac{1}{n-1}\,E\left[\left(\frac{y_i}{p_i} - Y\right) - \left(y'_c - Y\right) + \left(\frac{y'_i}{p_i} - \frac{Y}{p_i}\right) - \left(y'_{\text{HH}} - y'_c\right)\right]^2$$

(where y'_c is the value taken by y'_{HH} when the sample values y_i are known exactly, as in unistage cluster sampling)

$$= \frac{1}{n}\sum_{I=1}^{N} P_I\left(\frac{Y_I}{P_I} - Y\right)^2 + \frac{1}{n}\sum_{I=1}^{N} \frac{V_2\left(Y'_I\right)}{P_I} , \tag{5.2.6}$$

which is equivalent to the expression for $V\left(y'_{\text{HH}}\right)$ in (5.2.5). Hence

$$v\left(y'_{\text{HH}}\right) = \frac{1}{n(n-1)}\sum_{i=1}^{n}\left(\frac{y'_i}{p_i} - y'_{\text{HH}}\right)^2 \tag{5.2.7}$$

is an unbiased estimator of the total variance of y'_{HH}.

If it is desired to estimate components of variance for design purposes, an unbiased estimator of second and lower stage variance is clearly

$$v_2\left(y'_{\text{HH}}\right) = \frac{1}{n^2}\sum_{i=1}^{n} \frac{v_2\left(y'_i\right)}{p_i^2} , \tag{5.2.8}$$

where $v_2\left(y'_i\right)$ is an unbiased estimator of $V_2\left(y'_i\right)$. An unbiased estimator of first stage variance may be obtained by subtracting (5.2.8) from (5.2.7):

$$v_1\left(y'_{\text{HH}}\right) = \frac{1}{n(n-1)} \sum_{i=1}^{n} \left(\frac{y_i}{p_i} - y'_{\text{HH}}\right)^2 - \frac{1}{n^2} \sum_{i=1}^{n} \frac{v_2\left(y'_i\right)}{p_i^2} . \tag{5.2.9}$$

It was pointed out in Chapter 1 (Section 1.3) that the principal virtue of multinomial sampling with the Hansen-Hurwitz estimator lies in its simplicity and convenience. Expression (5.2.7) for the unbiased estimator of total variance is a further example of this virtue. However if they are used with a second and lower stage sample design other than that specified in this Subsection, (5.2.7) ceases to be an unbiased estimator and the principal motive for using this sampling strategy is largely undermined. A small compromise is, nevertheless, sometimes adopted. If the selection is such that, at the last stage, two (or three) independent *srswor* samples are required within a selected penultimate stage unit, a single *srswor* sample two (or three) times the size is selected instead. This reduces the variance of y'_{HH} slightly, but increases the value of expression (5.2.7), thus introducing a bias into $v\left(y'_{\text{HH}}\right)$. The magnitude of this bias is usually negligible.

5.2.2 General Formulae for Multistage Sampling Without Replacement

It is possible (but slightly more difficult) to derive the above results for the Hansen-Hurwitz estimator using general formulae developed by Rao (1975) for all multistage sampling schemes. Rao's analysis is, however, convenient to apply to without replacement schemes and will be used for that purpose here.

A general linear unbiased multistage estimator of Y may be written

$$y' = \sum_{I \in s} a_{Is} Y'_I = \sum_{I=1}^{N} a'_{Is} Y'_I , \tag{5.2.10}$$

where the a_{Is} are real numbers predetermined for each sample s , and

$$a'_{Is} = \begin{cases} a_{Is} & \text{if } I \in s , \\ 0 & \text{otherwise.} \end{cases}$$

The a_{Is} satisfy the unbiasedness condition that

$$E a_{Is} = \sum_{I \in s} a_{Is} p(s) = 1 , \tag{5.2.11}$$

where $p(s)$ is the probability of selecting the sample s . The variance of y' is

$$V(y') = V\left(y'_c\right) + E \sum_{I \in s} a_{Is}^2 V_{2s}\left(Y'_I\right) , \tag{5.2.12}$$

where $V_{2s}\left(Y'_I\right)$ is the conditional variance of Y'_I given s and

$$y_c' = \sum_{I=1}^{N} a_{Is}' Y_I \tag{5.2.13}$$

is the unbiased estimator of Y for unistage cluster sampling. We assume that an unbiased estimator can be formed for $V(y_c')$, namely

$$v(y_c') = \sum_{I\in s} b_{Is} Y_I^2 + \sum\sum_{\substack{I,J\in s\\ J>I}} d_{Is} Y_I Y_J , \tag{5.2.14}$$

where b_{Is} and d_{Is} are real numbers predetermined for every s. We further assume that an unbiased estimator $v_{2s}(Y_I')$ is available for given s, that is, that

$$Ev_{2s}(Y_I') = V_{2s}(Y_I') , \quad I \in s . \tag{5.2.15}$$

For the case where $V_{2s}(Y_I')$ is independent of s, and can therefore be written $V_2(Y_I')$, Raj (1966) proved that an unbiased estimator of $V(y')$ is

$$v_A(y') = \sum_{I\in s} b_{Is} Y_I'^2 + \sum\sum_{I,J\in s} d_{Is} Y_I' Y_J' + \sum_{I\in s} a_{Is} v_2(Y_I') . \tag{5.2.16}$$

An alternative unbiased estimator given by Rao (1975) and valid even when $V_{2s}(Y_I')$ is not independent of s is

$$v_B(y') = \sum_{I\in s} b_{Is} Y_I'^2 + \sum\sum_{\substack{I,J\in s\\ J>I}} d_{Is} Y_I' Y_J' + \sum_{I\in s} \left(a_{Is}^2 - b_{Is}\right) v_{2s}(Y_I') . \tag{5.2.17}$$

Quite apart from its more general application, $v_B(y')$ is preferable to $v_A(y')$ particularly when it comes to the estimation of variance components. Since the second stage component of variance is $\sum_{I\in s} a_{Is}^2 V_{2s}(Y_I')$, an unbiased estimator of the first stage component of variance derivable from (5.2.16) is

$$v_{A1}(y') = \sum_{I\in s} b_{Is} Y_I'^2 + \sum\sum_{\substack{I,J\in s\\ J>I}} d_{Is} Y_I' Y_J' + \sum_{I\in s} a_{Is} v_2(Y_I') - \sum_{I\in s} a_{Is}^2 v_2(Y_I') , \tag{5.2.18}$$

whereas that derivable from (5.2.17) is

$$v_{B1}(y') = \sum_{I\in s} b_{Is} Y_I'^2 + \sum\sum_{\substack{I,J\in s\\ J>I}} d_{Is} Y_I' Y_J' - \sum_{I\in s} b_{Is} v_{2s}(Y_I') . \tag{5.2.19}$$

The expression (5.2.19) is a more appropriate estimator of first stage variance because

$$Eb_{Is} Y_I'^2 = b_{Is} Y_I^2 + b_{Is} V_{2s}(Y_I') , \tag{5.2.20}$$

and the bias in the estimation of first stage variance caused by the use of Y'_I rather than Y_I is compensated term by term in (5.2.19) whereas in (5.2.18) it is only compensated by way of expectation over all first stage sample units. For further discussion of this point see Brewer and Hanif (1970).

5.2.3 Application to Particular Estimators

The following unbiased multistage variance estimators on the pattern of (5.2.17) were presented by Brewer and Hanif (1970).

A. Sen-Yates-Grundy Variance Estimator for the Horvitz-Thompson Estimator

$$v_{\text{SYG}}(y'_{\text{HT}}) = \sum_{\substack{I,J\ s \\ J>I}}\sum \left(\frac{\pi_I\pi_J-\pi_{IJ}}{\pi_{IJ}}\right)\left(\frac{Y'_I}{\pi_I}-\frac{Y'_J}{\pi_J}\right)^2 + \left\{\frac{1}{n-1}-\frac{\pi_I\pi_J-\pi_{IJ}}{\pi_{IJ}}\right\}\left\{\frac{v_{2s}(Y'_I)}{\pi_I^2}+\frac{v_{2s}(Y'_J)}{\pi_J^2}\right\} . \quad (5.2.21)$$

B. Raj's Variance Estimator for $n = 2$

(i) Ordered form

$$v(t_{\text{mean}}) = \tfrac{1}{4}\left[\left(1-p_1^2\right)\left(\frac{y'_1}{p_1}-\frac{y'_2}{p_2}\right)^2-(1-p_1)^2\left\{\frac{v_{2s}(y'_1)}{p_1^2}+\frac{v_{2s}(y'_2)}{p_2^2}\right\} + \left\{(1+p_1)^2\frac{v_{2s}(y'_1)}{p_1^2}+(1-p_1)^2\frac{v_{2s}(y'_2)}{p_2^2}\right\}\right] . \quad (5.2.22)$$

The terms $(1-p_1)^2 v_{2s}(y'_2)/p_2^2$ cancel, but have been retained above for clarity.

(ii) Unordered form

$$v_M(t_{\text{mean}}) = \tfrac{1}{4}\left[(1-p_1)(1-p_2)\left(\frac{y'_1}{p_1}-\frac{y'_2}{p_2}\right)^2-(1-p_1)(1-p_2)\left\{\frac{v_{2s}(y'_1)}{p_1^2}+\frac{v_{2s}(y'_2)}{p_2^2}\right\} + (1+p_1)^2\frac{v_{2s}(y'_1)}{p_1^2}+(1-p_1)^2\frac{v_{2s}(y'_2)}{p_2^2}\right] . \quad (5.2.23)$$

C. Murthy's Variance Estimator for $n = 2$

$$v(t_{\text{symm}}) = \frac{(1-p_1)(1-p_2)(1-p_1-p_2)}{(2-p_1-p_2)^2}\left(\frac{y_1'}{p_1} - \frac{y_2'}{p_2}\right)^2$$

$$- \frac{(1-p_1)(1-p_2)(1-p_1-p_2)}{(2-p_1-p_2)^2}\left(\frac{v_{2s}(y_1')}{p_1^2} + \frac{v_{2s}(y_2')}{p_2^2}\right)$$

$$+ \frac{(1-p_2)^2}{(2-p_1-p_2)^2}\frac{v_{2s}(y_1')}{p_1^2} + \frac{(1-p_1)^2}{(2-p_1-p_2)^2}\frac{v_{2s}(y_2')}{p_2^2} . \qquad (5.2.24)$$

D. RHC Variance Estimator

$$v(y'_{\text{RHC}}) = \left[\left(\sum_{i=1}^{n} N_i^2 - N\right) \div \left(N^2 - \sum_{i=1}^{n} N_i^2\right)\right] \sum_{i=1}^{n} \pi_i \left[\left(\frac{y'_{it}}{p_{it}} - y'_{\text{RHC}}\right)^2\right.$$

$$\left. - \left\{(1-\pi_i)^2 \frac{v_{2s}(y'_{it})}{p_{it}^2} + \sum_{\substack{j=1\\ j\neq i}}^{n} \pi_j^2 \frac{v_{2s}(y'_{jt})}{p_{jt}^2}\right\}\right] + \sum_{i=1}^{n} \pi_i^2 \frac{v_{2s}(y'_{it})}{p_{it}^2} . \qquad (5.2.25)$$

For the derivation of (5.2.25) see Hanif (1969), pp. 115-119.

The following additional formula may also be useful.

E. Approximate Variance Estimator (3.7.4) for Horvitz-Thompson Estimator

$$v_{\text{APP}}(y'_{\text{HT}}) = \left(1 - \left(\sum_{I=1}^{N} \pi_I^{2\gamma}\right) \div \left(\sum_{I=1}^{N} \pi_I^{2\gamma-1}\right)\right)\left[\frac{n}{n-1}\sum_{I\in s}\left(\frac{Y_I'}{\pi_I} - \frac{y'_{\text{HT}}}{n}\right)^2 - \sum_{I\in s}\frac{V_{2s}(Y_I')}{\pi_I^2}\right]$$

$$+ \sum_{I\in s}\frac{V_{2s}(Y_I')}{\pi_I^2} \qquad (5.2.26)$$

$$= \frac{n}{n-1}\left(1 - \left(\sum_{I=1}^{N} \pi_I^{2\gamma}\right) \div \left(\sum_{I=1}^{N} \pi_I^{2\gamma-1}\right)\right)\sum_{I\in s}\left(\frac{Y_I'}{\pi_I} - \frac{y'_{\text{HT}}}{n}\right)^2$$

$$+ \left(\sum_{I=1}^{N} \pi_I^{2\gamma}\right) \div \left(\sum_{I=1}^{N} \pi_I^{2\gamma-1}\right) \sum_{I\in s}\frac{V_{2s}(Y_I')}{\pi_I^2} . \qquad (5.2.27)$$

5.3. RATIO ESTIMATION IN MULTISTAGE SAMPLING

The analysis of this Chapter has so far been carried out in terms of unbiased estimation. The measures of size used in unequal probability sampling to some extent play the same role as benchmarks do in ratio estimation. For that reason there is sometimes little point in applying further ratio corrections. Frequently, however, such corrections are applied, either because a certain benchmark is particularly appropriate for some estimand variable (and while there can be only one measure of size used in selection, each estimand variable may have its own benchmark for ratio estimation) or else because some of the estimates are required to add to a specified total.

It has not been considered necessary to spell out the consequences of multistage ratio estimation in the context of a monograph on unequal probability sampling. The basic principles can be found in the text-books, for example, Cochran (1977), p. 317ff. Given the relevant formulae for the unbiased estimators from which the ratios are constructed, the application of these principles to particular cases is usually straightforward, see Hanif (1970a).

CHAPTER 6

AN OPTIMAL SAMPLING STRATEGY FOR LARGE UNISTAGE SAMPLES

6.1. INTRODUCTION

The sampling strategy to be described in this Chapter (Brewer, 1979) was devised for use in the context of large-scale surveys of populations containing units of very different sizes, such as official surveys of establishments and enterprises. Because the samples required are large, asymptotic theory is appropriate. However, the sample may be a significant proportion of the population, and consequently the finite population correction is allowed for.

The sampling strategy commonly used in such surveys is roughly as follows:

1. The population is stratified by type (industry, description, geographical location, and so on).
2. The type strata are further stratified by size. This is partly to ensure that each size of unit is adequately represented in the sample, and partly to take advantage of the markedly heteroscedastic nature of the population when it comes to the next step (sample allocation).
3. The sample is allocated over all strata in such a fashion as to ensure that the required degree of accuracy will be achieved for each estimate and that, subject to that requirement, the number of sample units is minimized. This minimization is achieved by Neyman allocation in which the sample fraction within a given size-type strata is proportional to the population standard deviation (psd) within it. (Typically, this psd

is not the simple psd of the estimand variable Y_I, but the psd of that variable adjusted for the variance of and covariance with some appropriate supplementary or benchmark variable Z_I used in the ratio estimation of Step 5.)

4. The number of sample units allocated to each particular type-size stratum is selected using *srswor*.

5. Within each type-size stratum, each estimand total Y_h is estimated using the conventional ratio estimator

$$y_h'' = \left[\left(\sum_{i=1}^{n_h} y_{hi}\right) \div \left(\sum_{i=1}^{n_h} z_{hi}\right)\right] Z_h \tag{6.1.1}$$

where Z_h is the total of the benchmark variable for the hth stratum. The use of an asymptotically unbiased rather than a strictly unbiased estimator is usually justified in this context because the scale of the survey ensures large sample sizes in each stratum. Some situations arise, however, where the number of strata is so large and the sample within each so small that combined ratio estimation is required in order to take advantage of this asymptotic property.

Provided the number of size strata is sufficient to pick up most of the advantages of differential sampling fractions, there is only one significant disadvantage in the above procedure. This is that size stratification inhibits the use of a more detailed industry or description stratification, which can often yield considerable rewards in terms of additional accuracy for a given sample size.

The only alternative procedures which the authors know to have been used on any scale in this context are *πpswor* selection with Horvitz-Thompson estimation, and nearly equivalent schemes. These enable size stratification to be side-stepped, but tend to give problems whenever very small units are selected, because of their high relative variabilities and the high sample weights which the Horvitz-Thompson estimator attaches to them.

The alternative procedure which will be presented in Section 6.2 is in one sense a compromise. It provides a sample weighted towards the larger units, but not to the extent that embarrassment occurs when small units are chosen. It retains (and even marginally improves) the accuracy of the size-stratified scheme. Most importantly, in abolishing the need for size stratification it opens the road to the use of finer industry or description stratification.

This alternative procedure uses the approach of Godambe (1955) and Godambe and Joshi (1965). A super-population model is postulated as providing a fair approximation to the actual situation. A class of estimators (or, more properly, predictors) of the population total is adopted on the basis of that model. The condition of (asymptotic) design unbiasedness is imposed, and the probabilities of selection are chosen to minimize the design expectation of the predictor's model based mean square error thus optimizing within the class.

In Section 6.3 some special cases of interest are examined.

In Section 6.4 an analogy is drawn between the traditional stratified random sampling schemes and the purely model-based robust schemes of Royall and Herson (1973a, b). On the basis of that analogy a modification to the Royall-Herson approach is suggested which also by-passes the need for stratification.

6.2. AN ALTERNATIVE PROCEDURE WITH A CONVENTIONAL SAMPLING RATIONALE

Recognizing that, given a sample containing n distinct units, there is an estimation problem only for $(N-n)$ values of Y_I not observed in the sample, the superpopulation model (1.8.5) implies that a sensible class of predictors for Y is

$$y^* = \sum_{I\in s} Y_I + \hat{\beta}\left(Z - \sum_{I\in s} Z_I\right) , \tag{6.2.1}$$

where $\hat{\beta}$ is a sample estimator of β . The best linear unbiased (BLU) estimator of β used in this context by Samiuddin, Hanif and Asad (1978) is

$$\hat{\beta}_{\mathrm{BLU}} = \sum_{I\in s} Y_I Z_I \sigma_I^{-2} \Big/ \sum_{I\in s} Z_I^2 \sigma_I^{-2} . \tag{6.2.2}$$

It will be seen later that the use of $\hat{\beta}_{\mathrm{BLU}}$ in (6.2.1) results in an estimator which is asymptotically unbiased over all possible samples (asymptotically design unbiased) only at the expense of having selection probabilities that yield comparatively inefficient samples. However, model (1.8.5) indicates that the dependence of Y_I on Z_I is linear and homogeneous. We therefore use the general homogeneous linear (and model unbiased) estimator

$$\hat{\beta} = \sum_{I\in s} W_I Y_I \Big/ \sum_{I\in s} W_I Z_I , \tag{6.2.3}$$

where the W_I are completely arbitrary. This ensures that y^* as defined by (6.2.1) and (6.2.3) is also unbiased under the model (1.8.5). We now require that this y^* be asymptotically unbiased over repeated sampling.

Since asymptotic analysis must allow the sample size, and hence the population size, to tend to infinity, it must be y^*/N rather than y^* which has the finite limit. Further it is necessary to define the manner in which the π_I behave as the sample and population sizes tend to infinity. This will be done as follows.

(1) The original population of N units is exactly reproduced $(k-1)$ times, yielding k populations of N units each in total.

(2) A sample is selected from each of these k populations in turn, using the same sample selection procedure (and hence the same π_I) for each one.

(3) The k populations are aggregated to a population of size $N_k = kN$ units with a population total $Y_k = kY$.

(4) The k samples are aggregated to a sample of n_k units.

(5) The estimator y^*_k of the population total Y_k is formed using equations (6.2.1) and (6.2.3).

(6) k is allowed to tend to infinity.

Since the design expectation of a ratio is asymptotically equal to the ratio of the design expectations, we may now write

$$\underset{k\to\infty}{E}\left(y_k^*/N_k\right) = N^{-1}\left[\sum_{I=1}^{N} \pi_I Y_I + \left\{\left(\sum_{I=1}^{N} \pi_I W_I Y_I\right) \div \left(\sum_{I=1}^{N} \pi_I W_I Z_I\right)\right\} \sum_{I=1}^{N} \left(1-\pi_I\right) Z_I\right] . \quad (6.2.4)$$

The expression on the right hand side of (6.2.4) is equal to Y/N (and hence y^* is asymptotically unbiased) if and only if the coefficient of each Y_I in the square bracket is unity; that is, if

$$\pi_I = \left\{\sum_{J=1}^{N} \pi_J W_J Z_J\right\} \div \left\{\sum_{J=1}^{N} \pi_J W_J Z_J + W_I \sum_{J=1}^{N} \left(1-\pi_J\right) Z_J\right\} = \frac{\alpha}{\alpha+W_I} , \quad (6.2.5)$$

where the value of α is proportional to the (arbitrary) sum of the W_I. With $\alpha = 1$ we obtain

$$W_I = \pi_I^{-1} - 1 . \quad (6.2.6)$$

We next seek to minimize the asymptotic value of the expectation (over all possible samples) of the variance of y_k^*/N_k under model (1.8.5). However since this is $O\left(k^{-1}\right)$ we minimize instead the limit of the expected variance of $y_k^*/N_k^{\frac{1}{2}}$. Using the same asymptotic argument as in the construction of (6.2.4),

$$\lim_{k\to\infty} EE^*\left\{\left(y_k^*-Y_k\right)^2/N_k\right\}$$

$$= N^{-1}\left[\left\{\left\{\sum_{I=1}^{N} \left(1-\pi_I\right) Z_I\right\}^2 \div \left\{\sum_{I=1}^{N} \pi_I W_I Z_I\right\}^2\right\} \sum_{I=1}^{N} \pi_I W_I^2 \sigma_I^2 + \sum_{I=1}^{N} \left(1-\pi_I\right) \sigma_I^2\right] . \quad (6.2.7)$$

Substituting (6.2.5) in (6.2.7), setting $\alpha = 1$ and simplifying,

$$\lim_{k\to\infty} EE^*\left\{\left(y_k^*-Y_k\right)^2/N_k\right\} = N^{-1}\sum_{I=1}^{N} W_I\sigma_I^2 . \tag{6.2.8}$$

By Schwartz's inequality,

$$\left\{\sum_{I=1}^{N}\sigma_I\right\}^2 \le \left\{\sum_{I=1}^{N}\left(1+W_I\right)^{-1}\right\}\left\{\sum_{I=1}^{N}\left(1+W_I\right)\sigma_I^2\right\}$$

$$= n\left\{\sum_{I=1}^{N}\sigma_I^2 + \sum_{I=1}^{N} W_I\sigma_I^2\right\} . \tag{6.2.9}$$

Hence

$$\sum_{I=1}^{N} W_I\sigma_I^2 \ge n^{-1}\left\{\sum_{I=1}^{N}\sigma_I\right\}^2 - \sum_{I=1}^{N}\sigma_I^2 . \tag{6.2.10}$$

The right hand side is achieved and expression (6.2.8) is minimized when

$$W_I = \left(n^{-1}\sigma_I^{-1}\sum_{J=1}^{N}\sigma_J^{-1}\right) , \tag{6.2.11}$$

and hence from (6.2.6) when

$$\pi_I = n\sigma_I \div \sum_{J=1}^{N}\sigma_J , \tag{6.2.12}$$

which is analogous to Neyman allocation.

Writing y^* in terms of the π_I we have

$$y^* = \sum_{I\in s} Y_I + \left(\sum_{I\in s} Y_I\left(\pi_I^{-1}-1\right) \div \sum_{I\in s} Z_I\left(\pi_I^{-1}-1\right)\right)\sum_{I\notin s} Z_I . \tag{6.2.13}$$

The expected variance of y^* when N is large may be written as $n^{-1}\sum_{I=1}^{N}\sigma_I^2\pi_I^{-1} - \sum_{I=1}^{N}\sigma_I^2$. This is equal to the asymptotic minimum of the expected variance of any design-unbiased estimator given those π_I , as shown by Godambe and Joshi (1965). Since the π_I values have also been optimized, the strategy consisting of y^* and $\pi_I \propto \sigma_I$ is asymptotically efficient.

6.3. SOME SPECIAL CASES

(i) If we have $\sigma_I^2 \propto Z_I^2$ so that $\pi_I = nZ_I/Z$, (6.2.13) simplifies, as does the generalized regression estimator of Cassel, Särndal and Wretman (CSW) (1976), to

$$y^* = \frac{Z}{n} \sum_{I \in s} \frac{Y_I}{Z_I} = \sum_{I \in s} \frac{Y_I}{\pi_I} , \tag{6.3.1}$$

the Horvitz-Thompson estimator with selection probabilities proportional to Z_I. If in addition the sample fraction is vanishingly small we have $\hat{\beta} = \hat{\beta}_{BLU}$. In all other cases it is preferable to settle for a less than optimum estimator of β in order to secure more favourable samples under the condition of asymptotic design unbiasedness.

(ii) If σ_I^2 is a constant for all I, $\pi_I = n/N$ and (6.2.13) becomes

$$y^* = \left(\sum_{I \in s} Y_I \div \sum_{I \in s} Z_I \right) Z , \tag{6.3.2}$$

the conventional ratio estimator with equal selection probabilities.

The CSW strategy in this case also uses equal selection probabilities, but the estimator is slightly unconventional. In basic form it corresponds to the conventional regression estimator, allowing for a non-zero intercept, but the slope parameter is estimated on the assumption that the intercept is zero.

(iii) With $\sigma_I^2 = \sigma_0^2 Z_I$ the selection is weighted somewhat towards the larger units $\left(\pi_I \propto Z_I^{\frac{1}{2}}\right)$ but there is no particularly simple form for the estimator. With this variance function the use of $\hat{\beta}_{BLU}$ would have led to the conventional ratio estimator for Y, equation (6.3.2), and this would in turn have resulted in the choice of an equal probability selection scheme. Now the expression (6.2.8), which we are seeking to minimize, takes the value

$$N^{-1}\left(\frac{N}{n} - 1\right) Z\sigma_0^2$$

using $\hat{\beta}_{BLU}$, and the value

$$N^{-1}\left\{n^{-1}\left(\sum_{I=1}^{N} Z_I^{\frac{1}{2}}\right)^2 - Z\right\}\sigma_0^2$$

using the $\hat{\beta}$ of (6.2.3). The proportionate reduction resulting from the use of $\hat{\beta}$ rather than $\hat{\beta}_{BLU}$ depends on the sampling fraction and the distribution of the Z_I. If the sampling fraction is small, this proportionate reduction is

$$\lambda = \left\{NZ - \left(\sum_{I=1}^{N} Z_I^{\frac{1}{2}}\right)^2\right\} \div NZ = \text{Var}\left(Z_I^{\frac{1}{2}}\right)/\bar{Z} . \tag{6.3.3}$$

Clearly the value of λ increases with the relative dispersion of the Z_I. For the uniform distribution between 0 and any positive upper limit, $\lambda = 1/9$. For the right tailed triangular distribution $\lambda = 11/75$. For the lognormal distribution

$\Lambda(\mu, \sigma^2)$, λ varies with σ as indicated in the following table, taken in part from Aitchison and Brown (1957), pp. 154-155.

σ	Coefficient of variation	λ
0.5	0.5329	0.061
1	1.3108	0.221
2	7.321	0.632

Most economic populations of the type considered here have coefficients of variation well in excess of unity, indicating that there would be appreciable loss of efficiency in the use of $\hat{\beta}_{\text{BLU}}$. In current practice this problem is avoided by the use of size stratification. It will be recalled, however, that the main practical advantage claimed for the alternative sampling scheme of Section 6.2 is that it removes the need for size stratification.

6.4. THE ROYALL-HERSON MODEL-BASED ROBUST PROCEDURE

Royall and Herson (RH) (1973a, 1973b) following an approach entirely antithetical to that of the previous Sections, sought to obtain robustness in model-based terms without reference to any samples other than the one actually selected. They chose a primary model identical with (1.8.5) except for the additional requirement that $\sigma_I^2 = \sigma_0^2 Z_I$. They then substituted the BLU estimate of β into expression (6.2.1) thereby ensuring that their y^* was simultaneously the conventional ratio estimator and a BLU predictor of Y . However, rather than impose any probability sampling properties (asymptotic or otherwise) on this predictor, they ensured robustness against possible polynomial terms in a more general population model by placing restrictions directly on the moments of the sample itself. (Provided a sample obeying these restrictions could be obtained, the manner of its choosing was irrelevant to their analysis.)

Specifically, they showed that their BLU predictor of Y - the conventional ratio estimator - remained unbiased and in most practical circumstances BLU in terms of the general polynomial model

$$Y_I = \sum_{j=0}^{J} \delta_j \beta_j z_I^j + e_I ,$$

$$E^* e_I = 0 , \quad E^* e_I e_J = \begin{cases} V(Z_I) & \text{if } J = 1 , \\ 0 & \text{otherwise,} \end{cases} \tag{6.4.1}$$

(where $\delta_j = 1$ if Y_I depends on Z_I^j and $\delta_j = 0$ otherwise) provided the sample and population moments obeyed the restrictions:

$$\left(\sum_{I\in s} Z_I^j\right) \div \left(\sum_{I\in s} Z_I\right) = \left(\sum_{I=1}^{N} Z_I^j\right) \div \left(\sum_{I=1}^{N} Z_I\right) \tag{6.4.2}$$

for all $j = 0, 2, 3, \ldots, J$ such that $\delta_j = 1$.

A sample obeying these restrictions was said to be *balanced* up to order J.

RH further pointed out that the restriction to polynomial terms in their model was not a necessary one. Any *independent* variables could be treated in the same way provided the sample was balanced on these variables.

Two practical problems were recognized. The first was that the construction of a balanced sample presented some difficulties. RH noted that 'simple random sampling will frequently produce a sample which is a fair approximation to a balanced sample', but warned that 'if, as happens with predictable frequency, the randomly selected sample is badly unbalanced, then the ratio estimator is biased. In such a case the fact that simple random sampling will produce balanced samples *on the average* is of little consolation' (RH's emphasis). They therefore recommended that if a simple random sample was found on inspection to be badly unbalanced in some variable suspected to be of potential importance, that sample should be rejected and another chosen. Effectively, this recommendation amounted to the use of restricted randomization from among all those samples which were approximately balanced in the moments of those independent variables against which the estimator was particularly desired to be robust.

The second practical problem recognized by RH was that balanced samples were distinctly less efficient, from the point of view of the model-based variance of the ratio estimator, than samples where the choice was to some extent tilted towards the larger units. RH's remedy (1973b) was to allow stratification on size $n_h/N_h \propto \bar{Z}_h$ effectively Neyman allocation given the RH variance function - together with the selection of a balanced sample within each stratum and the use of the conventional separate ratio estimator.

Despite the entirely different approach underlying this recommendation, the analogy between this scheme and the conventional stratified sampling scheme mentioned in Section 6.1 is obvious. The only operational differences are in the choice of the criterion for optimum allocation of sample between strata and in the use of restricted rather than unrestricted randomization to ensure that the sample chosen for each stratum is reasonably well balanced.

In Section 6.2 it was shown to be possible, using conventional sampling methods, to improve on the traditional strategy of size stratification, Neyman allocation and ratio estimation. This prompts the thought that analogous arguments in the RH spirit might result in a corresponding improvement to the RH recommendations. Such arguments are presented in the following Section.

6.5. AN ALTERNATIVE MODEL-BASED ROBUST PROCEDURE

As pointed out by Scott, Brewer, and Ho (1978), the variance function, $\sigma_I^2 = \sigma_0^2 Z_I$ used by RH is not usually as realistic as one which allows the variance to increase more rapidly with size. For the moment, however, no explicit variance function will be assumed, and the model (1.8.5) without further specification will be adopted as the primary model.

It will be recalled that RH's next step was to adopt the BLU predictor of Y under the assumed primary model. However, in the conventional theory of Section 6.2 the choice of the BLU predictor implied the choice of comparatively inefficient samples once the condition of asymptotic design unbiasedness had been imposed. Suspecting that something analogous will occur when robustness against alternative models is required, we retain as our estimator the y^* defined by (6.2.1) and (6.2.3).

The third step is to choose an alternative model against which we wish our estimator to be robust. Noting, with RH, that there is no need to restrict to polynomial terms for this purpose we adopt as our alternative model,

$$Y_I = \sum_{l=0}^{K} \beta_l X_{lI} + \eta_I \; , \tag{6.5.1}$$

where $E^*\eta_I = 0$.

Here the X_{lI} represent in principle any and every function of every potentially observable characteristic on unit I, other than Y_I and Z_I themselves. However, in practice it is necessary to keep K within reasonable bounds.

The condition for the y^* of (6.2.1) and (6.2.3) to be model unbiased (not, of course, BLU) under (6.5.1) is

$$\left(\sum_{I\in s} W_I X_{lI}\right) \div \left(\sum_{I\in s} W_I Z_I\right) = \left(\sum_{I\notin s} X_{lI}\right) \div \left(\sum_{I\notin s} Z_{lI}\right) \; ; \quad l = 0, 1, \ldots, K \; . \tag{6.5.2}$$

The ratio of the weighted sample moments of X_{lI} and Z_I must equal the ratio of these moments, unweighted, in the remainder of the population.

At this point the practical problem emerges; how can samples be chosen so as to satisfy (6.5.2), even for those few independent variables against whose influence we wish our estimator to be robust? Not surprisingly the solution is similar to that of

RH. In the same way as simple random sampling will frequently produce a sample which is a fair approximation to a balanced sample, sampling with the probabilities defined by (6.2.5) will frequently produce a sample which to a fair approximation obeys the conditions (6.5.2). In the context set out in Section 6.1, where the samples are comparatively large, it is not unfair to appeal to the asymptotic result obtained by taking expectations of each sum in (6.2.4) separately. However, if the sample is not large it is still reasonable to select using the π_I of (6.2.5), and if the more important of the restrictions (6.5.2) are badly broken by the sample first selected, to reject it and try again.

The observation that the use of the π_I of (6.2.5) will frequently produce a fair approximation to the type of sample needed also provides the necessary information about the relative efficiency or inefficiency of these resulting samples from the point of view of the mean square error of the predictor y^* under the primary model (1.8.5). The asymptotic argument is straightforward. For sufficiently large samples selected with these probabilities the variances based on the primary model can be made to accord with equation (6.2.8) arbitrarily closely. Restrictions on the randomization, by eliminating atypical samples, ensure that the limit is approached more rapidly than it would be without them. Hence asymptotically the model-based variance is minimized by the choice of the weights (6.2.11), in which case the probabilities of selection are given by (6.2.12). Although no such argument applies for small samples, these formulae can still be applied, and remain intuitively appealing.

The advantages of this scheme over the RH stratified sampling scheme correspond for the most part to the advantages of the Section 6.2 procedure over the traditional stratified sampling schemes; primarily in the abolition of the need for size stratification. Here, however, there is a further advantage. The RH stratified balanced sampling scheme prescribes $n_h/N_h \propto \bar{Z}_h$. This is a result dependent on the variance function of the RH primary model which has already been noted as insufficiently heteroscedastic. The scheme proposed in this Section allows for a more general variance function and in consequence permits a more efficient relationship between selection probability and unit size.

6.6. EFFICIENCY OF y^* AND SOME ALTERNATIVES

It has already been noted in Section 6.2 that the asymptotic expression for the expected variance of y^* is equal to the Godambe-Joshi bound for any design unbiased estimator. Using the standard Taylor series approximation the MSE of y^* can be shown to obey

$$V(y^*) \doteq V(y'_{HT}) + (\tilde{Y}^2/\tilde{Z}^2)V(z'_{HT}) - 2(\tilde{Y}/\tilde{Z})C(y'_{HT}, z'_{HT}) , \qquad (6.6.1)$$

where C denotes covariance and $y'_{\mathrm{HT}} = \sum_{I\in s} Y_I\pi_I^{-1}$, $z'_{\mathrm{HT}} = \sum_{I\in s} Z_I\pi_I^{-1}$,

$\tilde{Y} = \sum_{I=1}^{N} (1-\pi_I)Y_I$, and $\tilde{Z} = \sum_{I=1}^{N} (1-\pi_I)Z_I$. $\tilde{Y}$ and $\tilde{Z}$ are the expected values of the non-sample sum of the Y_I and Z_I respectively. But (6.6.1) is very similar to the mse of the Horvitz-Thompson ratio estimator,

$$y''_{\mathrm{HT}} = \frac{y'_{\mathrm{HT}}}{z'_{\mathrm{HT}}} Z , \qquad (6.6.2)$$

which obeys the approximation

$$V(y''_{\mathrm{HT}}) = V(y'_{\mathrm{HT}}) + \frac{Y^2}{Z^2} V(z'_{\mathrm{HT}}) - 2\frac{Y}{Z} C(y'_{\mathrm{HT}}, z'_{\mathrm{HT}}) . \qquad (6.6.3)$$

Comparing (6.6.1) with (6.6.3) it can further be shown that if y^* and y''_{HT} are more efficient than y'_{HT} , y^* will be more efficient than y''_{HT} when

$$\left(\sum_{I=1}^{N} \pi_I Y_I\right) \div \left(\sum_{I=1}^{N} \pi_I Z_I\right) > Y/Z . \qquad (6.6.4)$$

But under the primary model (1.8.5) the two sides of (6.6.4) are equal in expectation. It may therefore be concluded that there is no difference of importance between the efficiencies of y^* and y''_{HT} . As the latter is appreciably simpler and has been used by sampling statisticians for many years it seems unlikely that y^* will replace it in practice. Because of the basic similarity between y^* and y''_{HT} and the closeness of their mse formulae, this Chapter can perhaps best be regarded as a vindication of the use of y''_{HT} in current practice.

It is not yet, however, common to use y''_{HT} together with $\pi_I \propto \sigma_I$. This may be because of a reluctance to assume a model for the variance function, such as $\sigma_I^2 = \sigma^2 Z_I^{2\gamma}$. Perhaps now that it is clear that such a model need be used only for design purposes, and that the estimators y^* and y''_{HT} have a complete randomization-based rationale, such reluctance may be felt less strongly in future.

It should be noted, further, that y''_{HT} also satisfies approximately an RH-type robustness requirement analogous to balancing, in this case that

$$\left(\sum_{I\in s} X_{lI}\pi_I^{-1}\right) \div \left(\sum_{I\in s} Z_I\pi_I^{-1}\right) \doteq \left(\sum_{I=1}^{N} X_{lI}\right) \div \left(\sum_{I=1}^{N} Z_I\right) \qquad (6.6.5)$$

for all l .

A further alternative (or class of alternatives) is provided by the generalized regression estimator of Cassel, Särndal and Wretman (1976). In its most general form it may be written

$$y'_{\mathrm{CSW}} = y'_{\mathrm{HT}} + \hat{\beta}\left(Z - z'_{\mathrm{HT}}\right) , \qquad (6.6.6)$$

where $\hat{\beta}$ is essentially an estimator of the β of model (1.8.5). If we use the general homogeneous linear estimator of β given by (6.2.3) three interesting special cases emerge.

(1) If $W_I = \pi_I^{-1}$, $\hat{y}_{\mathrm{CSW}} = y''_{\mathrm{HT}}$.

(2) If $W_I = \left(\pi_I^{-1} - 1\right)$, $\hat{y}_{\mathrm{CSW}} = y^*$.

(3) If $W_I = Z_I \sigma_I^{-2}$, $\hat{\beta} = \hat{\beta}_{\mathrm{BLU}}$, and $\hat{y}_{\mathrm{CSW}}$ takes the form most favoured by its originators.

Now (6.6.6) is asymptotically unbiased and can therefore be regarded as design-robust regardless of the choice of $\hat{\beta}$. For this reason the use of $\hat{\beta}_{\mathrm{BLU}}$, which minimizes the expected variance of (6.6.6) under model (1.8.5), is particularly attractive.

The choice $W_I = \pi_I^{-1}$, however, is so close to $W_I = \left(\pi_I^{-1} - 1\right)$ that the differences in the model-based variances of y''_{HT} and y^* in actual samples are likely to be unimportant. In the absence of any empirical study, the inconclusive nature of (6.6.4) suggests that y^* , despite its theoretical attractions, has little to offer by way of real improvement over y''_{HT} . Asymptotically the expected variance of all three estimators tends to the Godambe-Joshi (1965) lower bound for any design unbiased estimator.

6.7. CHOICE OF SAMPLING PROCEDURE

The analysis of this Chapter so far has suggested the use of an estimator (either y^* or the very similar y''_{HT}) and a function to determine the π_I values (proportional to σ_I or Z_I). It has said nothing about the manner in which the sample should be selected given the π_I values.

With a sampling frame stratified only by type (of business, farm, factory, and so on) there is no call, as there is in area sampling, for limitation of the sample size to two units. In fact, with ratio estimation there is no need even for a fixed size sampling procedure. As was seen with the discussion of Poisson and collocated sampling in Chapter 4, the use of ratio estimation (even of a very simple kind) is sufficient to improve the efficiency of a non-fixed size sampling procedure virtually up to the standard of a fixed size procedure, provided only that the selection of an empty sample can be avoided.

Considerations of simplicity of selection, simplicity in variance estimation and, above all, rotatability and control, seem to the present authors to point to the choice of Poisson sampling as used by Ogus and Clark (1971). If it is felt necessary to reduce the probability of selecting an empty sample, or more generally to reduce the variability of the sample size, collocated sampling may be chosen instead. This would, however, involve some loss of simplicity in selection and variance estimation, and almost certainly some reduction in the stability of the variance estimator. There are also many exact *πpswor* procedures available for use with the Horvitz-Thompson estimator, which are compared in Chapter 3.

The ratio estimator y''_{HT} of (6.6.2) is not in general the same as that considered for Poisson and collocated sampling in Chapter 4. They are identical only in the special case where the π_I are proportional to Z_I (the optimal choice when $\gamma = 1$). In this instance, because of the non-fixed sample size and of the special term which relates to the possiblity of selecting an empty sample, the application of conventional ratio estimation theory is not entirely straightforward, and the results will be presented here.

The ratio estimator (6.6.2) now becomes

$$y''_{\text{HT}} = \left.\begin{array}{ll} \dfrac{y'_{\text{HT}}}{z'_{\text{HT}}} Z & \text{if } m > 0 \\ \\ = 0 & \text{otherwise} \end{array}\right\} \qquad (6.7.1)$$

Its mean square error obeys the approximation

$$V(y''_{\text{HT}}) \doteq V(y'_{\text{HT}}) + \frac{Y^2}{Z^2} V(z'_{\text{HT}}) - 2\frac{Y}{Z} C(y'_{\text{HT}}, z'_{\text{HT}}) + P_{S_0} Y^2 , \qquad (6.7.2)$$

where $P_{S_0} = P_0$ for Poisson and P_{0C} for collocated sampling and

$$C(y'_{\text{HT}}, z'_{\text{HT}}) = \sum_{I=1}^{N} \frac{(1-\pi_I)}{\pi_I} Y_I Z_I + \sum_{\substack{I,J=1 \\ J\neq I}}^{N}\sum \frac{\pi_{IJ}-\pi_I\pi_J}{\pi_I\pi_J} Y_I Z_J , \qquad (6.7.3)$$

the two variance terms being formed analogously. (For Poisson sampling $\pi_{IJ} = \pi_I\pi_J$ and the terms in $Y_I Z_J$, $J \neq I$, disappear.) An unbiased estimator of (6.7.3) is

$$c_1\left(y'_{HT}, z'_{HT}\right) = \sum_{I\in s} \frac{\left(1-\pi_I\right)}{\pi_I^2} Y_I Z_I + \sum\sum_{\substack{I,J\in s \\ J\neq I}} \frac{\pi_{IJ}-\pi_I\pi_J}{\pi_{IJ}\pi_I\pi_J} Y_I Z_J . \qquad (6.7.4)$$

However because of the non-fixed sample size a more stable variance estimator of (6.7.3) is

$$c_2\left(y'_{HT}, z'_{HT}\right) = \frac{n}{m} \sum_{I\in s} \frac{\left(1-\pi_I\right)}{\pi_I^2} Y_I Z_I + \frac{n(n-1)}{m(m-1)} \sum\sum_{\substack{I,J\in s \\ J\neq I}} \frac{\pi_{IJ}-\pi_I\pi_J}{\pi_{IJ}\pi_I\pi_J} Y_I Z_J . \qquad (6.7.5)$$

Analogous variance estimators may be formed for $V\left(y'_{HT}\right)$, $V\left(z'_{HT}\right)$, and the mean square error (6.7.2) is then estimated by

$$v\left(y''_{HT}\right) = v_2\left(y'_{HT}\right) + \left(\frac{y'_{HT}}{z'_{HT}}\right)^2 v_2\left(z'_{HT}\right) - 2\,\frac{y'_{HT}}{z'_{HT}}\,c_2\left(y'_{HT}, z'_{HT}\right) + P_{S_0} y''^2_{HT} . \qquad (6.7.6)$$

For y^* a possible variance estimator is

$$v(y^*) = v_2\left(y'_{HT}\right) + \hat{\beta}^2 v_2\left(z'_{HT}\right) - 2\hat{\beta} c_2\left(y'_{HT}, z'_{HT}\right) + P_{S_0} y^{*2} . \qquad (6.7.7)$$

For $\hat{y}_{CSW}$ no substantial improvement is possible over the variance estimator (6.7.6) already suggested for y''_{HT}. This is because formula (6.7.2) which provides the basis for estimating the variance of y''_{HT} is also the exact variance for $\hat{y}_{CSW}$ with Y/Z replacing $\hat{\beta}_{BLU}$. Additional terms in the variance formula reflecting the variance of $\hat{\beta}_{BLU}$ as an estimator of Y/Z are of a lower order of magnitude which has generally been ignored in ratio estimation.

APPENDIX A

WORKING PROBABILITIES FOR NARAIN'S AND FELLEGI'S PROCEDURES FOR $n = 2$

In this Appendix BASIC language routines are given for calculating the working probabilities for Narain's Procedure 7 and Fellegi's Procedure 13 for $n = 2$ based on the geometric properties of the problems described by Brewer and Undy (1962) and Brewer (1967).

In each of these Procedures it is necessary to find the unique real solution of an equation $f(T) = 0$ where $f(T)$ involves N square root terms, $\pm D_I$, one for each population unit (see equations (A.4), (A.5), (A.11) and (A.12) below). This unique real solution is known to require all the D_I , with the possible exception of that relating to the largest unit, D_N , to take a negative sign. Ignoring those parts of the curve where the other D_I take negative signs, there remain two branches of $f(T)$, the upper in which D_N takes the positive sign and the lower in which it takes the negative. These two branches meet at $T_{\min}$, the value of T for which $D_N = 0$.

The form of $f(T)$ is roughly hyperbolic. The slope of the lower branch is always negative and tends to a finite limit. The slope of the upper branch is $+\infty$ at $T = T_{\min}$, but the branch itself passes through a maximum and then decreases to a positive asymptote.

The algorithm commences by taking as its initial total value $T^{(1)} = T_{\min}$ (see equations (A.6) and (A.13) below). If $f\left(T_{\min}\right)$ is positive, the solution lies on the lower branch, and *vice versa*, and only the relevant branch is examined thereafter. If $T_{\min}$ is appreciably less than unity, the second value taken is $T^{(2)} = 1$, but otherwise it is assumed that the shape of the curve near $T = T_{\min}$ is sufficiently close to a parabola for it to be possible to find an approximate solution based on the variations in D_N only (see equations (A.7) and (A.14)). This ensures that $T^{(2)}$ is always real and greater than $T^{(1)}$.

For all subsequent iterations the standard practice is to set the next trial value $T^{(r+1)}$ by

$$T^{(r+1)} = T^{(r)} - f\left(T^{(r)}\right)/f'\left(T^{(r)}\right) , \tag{A.1}$$

where $f'\left(T^{(r)}\right)$ is the slope of $f(T)$ evaluated at $T^{(r)}$. Two departures from this standard are necessary. The first is if $T^{(r)}$ is less than $T^{(1)}$. In this case the algorithm brings the trial value closer to $T^{(2)}$ by putting

$$T^{(r+1)} = \left(T^{(r)}+T^{(2)}\right)/2 , \tag{A.2}$$

and testing whether $T^{(r+1)}$ is less than $T^{(1)}$. The second is if the solution lies on the upper branch and the value of $T^{(r)}$ is greater than that corresponding to the maximum of $f(T)$, that is, if $f\left(T^{(1)}\right)$ and $f'\left(T^{(r)}\right)$ are both negative. In this case the algorithm brings the trial value closer to $T^{(1)}$ by putting

$$T^{(r+1)} = \left(T^{(r)}+T^{(1)}\right)/2 \tag{A.3}$$

and testing again.

The essential equations are

(a) Narain's Procedure 7

$$2f(T) = (1+T)(N-2) - \sum_{I=1}^{N-1} D_I \pm D_N , \tag{A.4}$$

$$D_I = +\left\{\left(1+T-2P_I\right)^2-8P_I\right\}^{\frac{1}{2}} , \tag{A.5}$$

$$T^{(1)} = T_{\min} = 2P_N + \left(8P_N\right)^{\frac{1}{2}} - 1 , \tag{A.6}$$

$$\begin{aligned} T^{(2)} &= T^{(1)} + \left\{f\left(T^{(1)}\right)\right\}^2/\left(2P_N\right)^{\frac{1}{2}} \quad \text{if} \quad T^{(1)} > 0.99 \\ &= 1 \quad \text{otherwise,} \end{aligned} \tag{A.7}$$

$$2f'(T) = N - 2 - \sum_{I=1}^{N-1} \left(1+T-2P_I\right)/D_I \pm \left(1+T-2P_N\right)/D_N \ , \tag{A.8}$$

$$\begin{aligned} T_I &= 1 + T - 2P_I - D_I \ , \quad I < N \\ &= 1 + T - 2P_I \pm D_N \ , \quad I = N \ , \end{aligned} \tag{A.9}$$

$$P_I(1) = P_I(2) = T_I/\left(1+T_I\right) \ . \tag{A.10}$$

(b) Fellegi's Procedure 13

$$2f(T) = T(N-2) - \sum_{I=1}^{N-1} D_I \pm D_N \ , \tag{A.11}$$

$$D_I = +\left\{T\left(T-4P_I\right)\right\}^{\frac{1}{2}} \ , \tag{A.12}$$

$$T^{(1)} = T_{\min} = 4P_N \ , \tag{A.13}$$

$$\begin{aligned} T^{(2)} &= T^{(1)} + \left\{f\left(T^{(1)}\right)\right\}^2/P_N \quad \text{if} \quad T^{(1)} > 0.99 \\ &= 1 \quad \text{otherwise,} \end{aligned} \tag{A.14}$$

$$2f'(T) = N - 2 - \sum_{I=1}^{N-1} \left(T-2P_I\right)/D_I \pm \left(T-2P_N\right)D_N \ , \tag{A.15}$$

$$\begin{aligned} T_I &= \left(T-D_I\right)/2 \ , \quad I < N \\ &= \left(T\pm D_N\right)/2 \ , \quad I = N \ , \end{aligned} \tag{A.16}$$

$$P_I(2) = 1 - P_I/T_I \ . \tag{A.17}$$

The following BASIC programs are for the most part identical. Where they differ, the left hand column relates to Narain's Procedure 7 and the right hand column to Fellegi's Procedure 13.

```
10 REM NARAIN'S WKG PROBS                    10 FELLEGI'S WKG PROBS
                         20 DIM  P(50)
                         30 DIM  D(50)
                         40 DIM D1(50)
                         50 DIM T9(50)
60 DIM P1(50)                                 60 DIM P2(50)
                 70 OPEN "LP:" FOR OUTPUT AS FILE #1
                 80 INPUT N
                 90 PRINT #1:"N";N
                100 FOR J=1 TO N
                110 INPUT P(J)
                120 PRINT #1:"P(";J;")";P(J)
                130 NEXT J
                135 REM MINIMUM T
```

```
140 LET T1=2*P(N)+(8*P(N))↑.5-1                    140 LET T1=4*P(N)
150 PRINT #1:"T1";T
155 REM FIRST ITERATION
160 LET S=0
170 LET S1=0
180 FOR J=1 TO N-1
190 LET D(J)=((1+T1-2*P(J))↑2-8*P(J))↑.5           190 LET D(J)=(T1*(T1-4*P(N)))↑.5
200 PRINT #1:"D(";J;")";D(J)
210 LET D1(J)=(1+T1-2*P(J))/D(J)                   210 LET D1(J)=(T1-2*P(J))/D(J)
220 PRINT #1:"D1(";J;")";D1(J)
230 LET S=S+D(J)
240 LET S1=S1+D1(J)
250 NEXT J
260 PRINT #1:"S";S
270 PRINT #1:"S1";S1
280 LET F1=.5*((1+T1)*(N-2)-S)                     280 LET F1=.5*(T1*(N-2)-S)
290 PRINT #1:"F1";F1
291 REM SECOND ITERATION
292 IF T1>.99 THEN 300
294 T=1
296 TO TO 310
300 LET T=T1+F1↑2/(2*P(N))↑.5                      300 LET T=T1+F1↑2/P(N)
310 PRINT #1:"T2";T
312 T2=T
315 REM SUBSEQUENT ITERATIONS
320 FOR I=1 TO 20
330 LET S=0
340 LET S1=0
350 FOR J=1 TO N
360 LET D(J)=((1+T-2*P(J))↑2-8*P(J))↑.5            360 LET D(J)=(T*(T=4*P(J)))↑.5
370 PRINT #1:"D(";J;")";D(J)
380 LET D1(J)=(1+T-2*P(J))/D(J)                    380 LET D1(J)=(T-2*P(J))/D(J)
390 PRINT #1:"D1(";J;")";D1(J)
400 LET S=S+D(J)
410 LET S1=S1+D1(J)
420 NEXT J
425 REM CORRECTION FOR UPPER BRANCH
430 IF F1>0 THEN 460
440 LET S=S-2*D(N)
450 LET S1=S1-2*D1(N)
460 PRINT #1:"S";S
470 PRINT #1:"S1";S1
475 REM EVALUATION OF F
```

```
480 LET F=.5*((1+T)*(N-2)-S)                          480 LET F=15*(T*(N-2)-S)
                          485 PRINT #1:"F";F
                          487 REM IS UPPER BRANCH SLOPE -VE?
                          490 IF F1>0 THEN 540
                          500 IF (N-2-S1)>0 THEN 540
                          510 LET T=(T+T1)/2
                          520 PRINT #1:"T";T
                          530 GL TO 330
                          535 REM STANDARD ITERATION
                          540 LET T=T-2*F/(N-2-S1)
                          550 PRINT #1:"T";T
                          555 REM EXIT IF CONVERGED
                          560 IF ABS(F)<1E-6 THEN 580
                          562 REM IS T<T1?
                          564 IF T>T1 THEN 570
                          566 LET T=(T+T2)/2
                          567 PRINT #1:"T";T
                          568 GO TO 564
                          570 NEXT I
                          575 REM FINAL CALCULATIONS
                          580 LET C=0
                          590 FOR J=1 TO N

600 LET T9(J)=.5*(1+T-2*P(J)-D(J))                    600 LET T9(J)=.5*(T-D(J))
                       605 REM CORRECTION FOR UPPER BRANCH
                       610 IF J<N THEN 640
                       620 IF F1>0 THEN 640
                       630 LET T9(N)=T9(N)+D(N)

640 LET P1(J)=T9(J)/(1+T9(J))                         640 LET P2(J)=1-P(J)/T9(J)

650 PRINT #1:"P1(";J;")";P1(J)                        650 PRINT #1:"P2(";J;")";P2(J)

660 LET C=C+P1(J)                                     660 LET C=C+P2(J)
                                  670 NEXT J

680 PRINT #1:"CHECK SUM P1";C                         680 PRINT #1:"CHECK SUM P2";C
                                  690 CLOSE #1
                                  700 STOP
                                  710 END
```

The rate of convergence is satisfactory for both algorithms, being slowest when P_N is nearly its maximum value of 0.5 . The first few iterations after the second then decrease the value of $f(T)$ only by a factor of two, but the rate of convergence thereafter increases.

APPENDIX B

MEAN SQUARE ERROR OF THE ESTIMATOR y'' FOR POISSON AND COLLOCATED SAMPLING

Regardless of the procedure used to select the sample, the relative variances of y' and m, and their relative covariance are, respectively,

$$V_{y'}^2 = \frac{E(y'-Y)^2}{Y^2} = \frac{1}{Y^2}\left[\sum_{I=1}^{N}(1-\pi_I)\frac{Y_I^2}{\pi_I} + \sum_{\substack{I,J=1\\J\neq I}}^{N}\sum(\pi_{IJ}-\pi_I\pi_J)\frac{Y_I}{\pi_I}\frac{Y_J}{\pi_J}\right], \tag{B.1}$$

$$V_m^2 = \frac{E(m-n)^2}{n^2} = \frac{1}{n^2}\left[\sum_{I=1}^{N}(1-\pi_I)\pi_I + \sum_{\substack{I,J=1\\J\neq I}}^{N}\sum(\pi_{IJ}-\pi_I\pi_J)\right] \tag{B.2}$$

and

$$V_{y',m} = \frac{E(y'-Y)(m-n)}{nY} = \frac{1}{nY}\left[\sum_{I=1}^{N}(1-\pi_I)Y_I + \sum_{\substack{I,J=1\\J\neq I}}^{N}\sum(\pi_{IJ}-\pi_I\pi_J)\frac{Y_I}{\pi_I}\right].$$

Clearly

$$V_{y',m} = \frac{1}{nY}\left[2\sum_{I=1}^{N}(1-\pi_I)Y_I + \sum_{\substack{I,J=1\\J\neq I}}^{N}\sum(\pi_{IJ}-\pi_I\pi_J)\left(\frac{Y_I}{\pi_I}+\frac{Y_J}{\pi_J}\right)\right]. \tag{B.3}$$

Now let S be the set of possible samples, s_0 the null sample and P_s the probability that the sth sample is selected. Then y', m and y'' are random variables defined on S. Now we construct random variables z', ν and z'' on

$S - \{s_0\}$ by choosing s within $S - \{s_0\}$ with probability $P_s\left(1-P_{s_0}\right)^{-1}$ and defining $z'(s) = y'(s)$, $\nu(s) = m(s)$ and $z''(s) = n \cdot \frac{z'(s)}{\nu(s)} = y''(s)$. Then

$$E(z'-Y)^2 = \sum_{s\in\left(S-s_0\right)} \frac{P_s}{1-P_{s_0}} \left(y'(s)-Y\right)^2$$

$$= \left(1-P_{s_0}\right)^{-1}\left[\sum_{s\in S} P_s\left(y'(s)-Y\right)^2 - P_{s_0}Y^2\right] \quad \text{since } y'\left(s_0\right) = 0$$

$$= \left(1-P_{s_0}\right)^{-1}\left[E(y'-Y)^2 - P_{s_0}Y^2\right]$$

so

$$V^2_{z'} = \frac{E(z'-Y)^2}{Y^2} = \left(1-P_{s_0}\right)^{-1}\left(V^2_{y'}-P_{s_0}\right) .$$

Similarly

$$V^2_\nu = \left(1-P_{s_0}\right)^{-1}\left(V^2_m - P_{s_0}\right) \tag{B.4}$$

and

$$V_{z',\nu} = \left(1-P_{s_0}\right)^{-1}\left(V_{y',m} - P_{s_0}\right) .$$

But the mean square error of the ratio estimator y'' is approximately given by

$$Y^2V(y'') \doteq E(y''-Y)^2 = \sum_{s\in S} P_s\left(y''(s)-Y\right)^2$$

$$= P_{s_0}Y^2 + \left(1-P_{s_0}\right)E(z''-Y)^2$$

$$= P_{s_0}Y^2 + \left(1-P_{s_0}\right)Y^2V^2_{z''} . \tag{B.5}$$

But $V^2_{z''} \doteq V^2_{z'} - 2V_{z',\nu} + V^2_\nu$ since $\nu > 0$ always, and $\nu < 2n$ with high probability. Hence from (B.4) we obtain

$$V^2_{z''} \doteq \left(1-P_{s_0}\right)^{-1}\left[V^2_{y'} - 2V_{y',m} + V^2_m\right] , \tag{B.6}$$

and substituting the results from (B.1), (B.2), (B.3) and (B.6) into (B.5),

$$V(y'') \doteq P_{s_0}Y^2 + \sum_{I=1}^{N} \pi_I\left(1-\pi_I\right)\left(\frac{Y_I}{\pi_I} - \frac{Y}{n}\right)^2 + 2\sum_{J>I}^{N}\left(\pi_{IJ}-\pi_I\pi_J\right)\left(\frac{Y_I}{\pi_I} - \frac{Y}{n}\right)\left(\frac{Y_J}{\pi_J} - \frac{Y}{n}\right) . \tag{B.7}$$

Since for Poisson sampling $\pi_{IJ} = \pi_I\pi_J$ and $P_{s_0} = P_0$.

$$V\left(y''_{\mathrm{PS}}\right) = P_0Y^2 + \sum_{I=1}^{N} \pi_I\left(1-\pi_I\right)\left(\frac{Y_I}{\pi_I} - \frac{Y}{n}\right)^2 , \tag{4.2.26}$$

and the expression (4.2.36) for collocated sampling follows directly from (B.7) writing P_{0C} for P_{s_0} .

APPENDIX C

OVERLAP BETWEEN POISSON AND COLLOCATED SAMPLES

In Chapter 4 it was noted that Poisson and collocated samples could be controlled and related in a simple fashion using diagrams in which the inclusion probabilities π_I, π'_I were represented on the vertical axis and the random numbers r_I on the horizontal axis. Two such diagrams are given in Figures C.1 and C.2. Both figures indicate the relevant ranges of r_I values corresponding to inclusion in an original sample (selected with probabilities π_I) and a revised sample (selected with probabilities π'_I). Figure C.1 depicts the situation where the Ith unit is included in the original sample if $r_I < \max(0, \pi'_I-1+c)$ or $c \le r_I < \min(\pi_I+c, 1)$, and in the revised sample if $r_I < \max(0, \pi_I-1+c+d)$ or $c + d \le r_I < \min(\pi'_I+c+d, 1)$. In Figure C.2 the value of c has been set equal to zero. Since this simplification can be achieved without loss of generality, the remainder of this Appendix will relate to the situation depicted in Figure C.2.

It should be noted that in both diagrams the inclusion regions for the original sample relate to the inclusion probabilities π_I and those for the revised sample relate to the inclusion probabilities π'_I. Thus, for example, the Ith unit will be included in both samples if and only if the pair $\{\pi_I, r_I\}$ is in a region marked "IN ORIGINAL" or "IN BOTH SAMPLES" and the pair $\{\pi'_I, r_I\}$ is in a region marked "IN REVISED" or "IN BOTH SAMPLES".

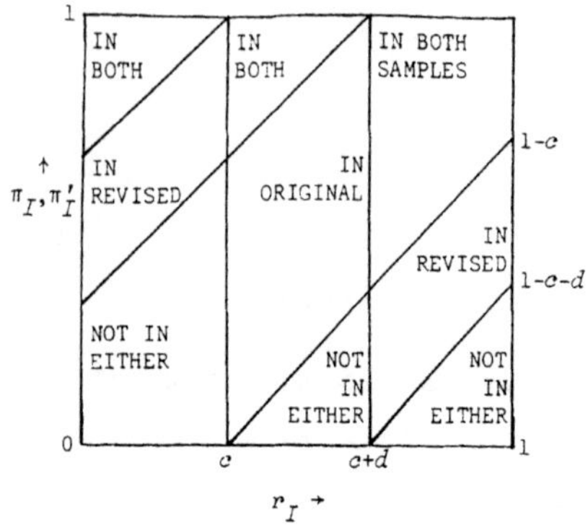

FIGURE C.1

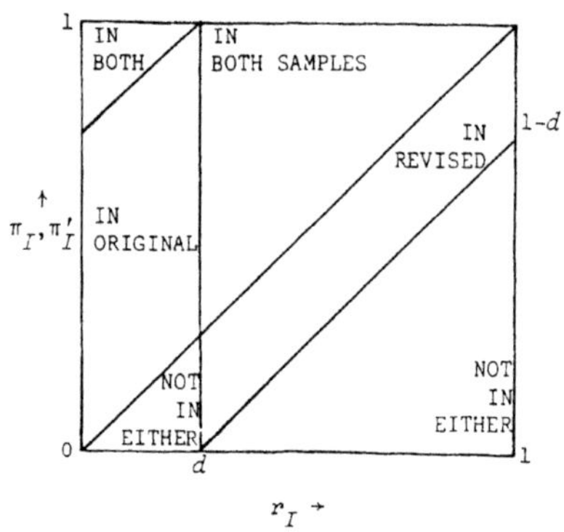

FIGURE C.2

There are four cases in all, which can most easily be considered in two pairs.

Consider first the situation where $\pi'_I \leq 1-d$. Then the Ith unit is in the original sample if $r_I < \pi_I$ and in the revised if $d \leq r_I < \pi'_I + d$. It is therefore in both if $d \leq r_I < \min(\pi_I, \pi'_I+d)$.

CASE 1: $\pi'_I \leq 1-d$ and $\pi_I \leq d$.

Since $\pi_I \leq d$ the unit cannot be in both samples and $B_I = 0$.

Case 2: $\pi_I' \le 1-d$ and $\pi_I > d$.

In this case both π_I and $\pi_I' + d$ exceed d , and $B_I = \min(\pi_I - d, \pi_I')$.

Now consider the situation where $\pi_I' > 1-d$. Then the Ith unit is in the original sample if $r_I < \pi_I$ and in the revised if $d \le r_I < \pi_I' + d$ or $r_I < \pi_I' - 1 + d$. It is therefore in both if $d \le r_I < \min(\pi_I, \pi_I'+d)$ or $r_I < \min(\pi_I, \pi_I'-1+d)$.

Case 3: $\pi_I' > 1-d$ and $\pi_I \le d$.

Since $\pi_I \le d$ the first condition cannot be fulfilled, and hence $B_I = \min(\pi_I, \pi_I'-1+d)$.

Case 4: $\pi_I' > 1-d$ and $\pi_I > d$.

In this case $\min(\pi_I, \pi_I'+d) = \pi_I$ and the first condition occurs with probability $\pi_I - d$. Also $\min(\pi_I, \pi_I'-1+d) = \pi_I' - 1 + d$ so the second condition holds with probability $\pi_I' - 1 + d$. But the two conditions are mutually exclusive, hence $B_I = (\pi_I - d) + (\pi_I' - 1 + d) = \pi_I + \pi_I' - 1$. The four possible formulae for B_I are summarized in tabular form in (4.9.1).

Referring to this table we can describe $B_I(d)$ as a function in d . In the general case B_I assumes its maximum value of $\min(\pi_I, \pi_I')$ on $[0, \max(0, \pi_I - \pi_I')]$ and $[\min(1, 1-\pi_I'+\pi_I), 1]$, decreases with gradient -1 over the interval $[\max(0, \pi_I - \pi_I'), \min(\pi_I, 1-\pi_I')]$, assumes its minimum value of $\max(0, \pi_I'+\pi_I-1)$ on $[\min(\pi_I, 1-\pi_I'), \max(\pi_I, 1-\pi_I')]$ and increases with gradient 1 over $[\max(\pi_I, 1-\pi_I'), \min(1, 1-\pi_I'+\pi_I)]$. To maximise the expected overlap between the two samples we choose $d = 0$, or equivalently

$$\sum_{I=1}^{N} B_I(0) = \sum_{I=1}^{N} B_I(1) = \sum_{I=1}^{N} \max_d B_I(d) = \sum_{I=1}^{N} \min(\pi_I, \pi_I') .$$

Now $\min B_I(d) = B_I(\tfrac12)$, unless π_I and $1 - \pi_I'$ are either both less than $\tfrac12$ or both greater than $\tfrac12$. If $\pi_I, 1-\pi_I' < \tfrac12$, $B_I(\tfrac12) = \min(\pi_I' - \tfrac12, \pi_I)$ and so

$$B_I(\tfrac12) - \min B_I(d) = \min(\pi_I' - \tfrac12, \pi_I) - \max(0, \pi_I + \pi_I' - 1) \le \pi_I' - \tfrac12 .$$

Similarly if $\pi_I, 1-\pi_I' > \tfrac12$, we have

$$B_I(\tfrac12) - \min B_I(d) \le \pi_I - \tfrac12 .$$

Hence

$$\sum_{I}^{N} B_I(\tfrac{1}{2}) - \min \sum_{I}^{N} B_I(d) \le \sum_{I}^{N} \{B_I(\tfrac{1}{2}) - \min B_I(d)\}$$

$$\le \sum_{\pi_I, 1-\pi_I'>\frac{1}{2}} (\pi_I - \tfrac{1}{2}) + \sum_{\pi_I, 1-\pi_I'<\frac{1}{2}} (\pi_I' - \tfrac{1}{2}) .$$

In practice there will be few cases where π_I and $1 - \pi_I'$ are both less than $\frac{1}{2}$ or both greater than $\frac{1}{2}$ and in such cases the respective values $\pi_I - \frac{1}{2}$ and $\pi_I' - \frac{1}{2}$, being less than $|\pi_I - \pi_I'|$, will usually be small. Hence taking $d = \frac{1}{2}$ should be a good approximation to minimizing the expected overlap between two samples, and this minimum will be approximated by

$$\sum_{I=1}^{N} \min B_I(d) \doteq \sum_{\pi_I+\pi_I'>1} (\pi_I' + \pi_I - 1) .$$

APPENDIX D

CALCULATION OF π_{IJ} FOR COLLOCATED SAMPLING

It can be shown that π_{IJ} is independent of c , so we shall assume $c = 0$. The Ith unit is in sample iff $r_I < \pi_I$ which, in terms of our specification of r_I , reduces to $L_I+\gamma < N\pi_I+1$. Defining $K_I = N\pi_I - [N\pi_I]$ where $[N\pi_I]$ is the integral part of $N\pi_I$, we can express this condition in the following form:

The Ith unit is in sample iff either

$$L_I \le [N\pi_I] + 1 \quad \text{and} \quad \gamma < K_I$$

or else

$$L_I \le [N\pi_I] \quad \text{and} \quad \gamma \ge K_I \ . \qquad (D.1)$$

Assume without loss of generality that $\pi_J \ge \pi_I$ and hence that $[N\pi_J] \ge [N\pi_I]$. There are two cases to consider:

Case (i). If $K_I \le K_J$, then referring to (D.1) and noting the independence of γ and L_I we have

$$\pi_{IJ} = \Pr\{\gamma < K_I;\ L_I \le [N\pi_I]+1;\ L_J \le [N\pi_J]+1 \mid \pi_J \ge \pi_I\}$$
$$+ \Pr\{K_I \le \gamma < K_J;\ L_I \le [N\pi_I];\ L_J \le [N\pi_J]+1 \mid \pi_J \ge \pi_I\}$$
$$+ \Pr\{\gamma \ge K_J;\ L_I \le [N\pi_I];\ L_J \le [N\pi_J] \mid \pi_J \ge \pi_I\}$$
$$= \frac{1}{N(N-1)} \{K_I([N\pi_I]+1)[N\pi_J]+(K_J-K_I)[N\pi_I][N\pi_J]+(1-K_J)[N\pi_I]([N\pi_J]-1)\}$$
$$= \frac{1}{N(N-1)} \{[N\pi_I](N\pi_J-1)+K_I[N\pi_J]\} \ . \qquad (D.2)$$

Case (ii). If $K_I > K_J$ we calculate π_{IJ} as above except that the conditions on γ now become $\gamma < K_J$, $K_J \le \gamma < K_I$ and $\gamma \ge K_I$ respectively, and the requirements on L_I, L_J are changed accordingly. We thus obtain,

$$\pi_{IJ} = \begin{cases} \frac{1}{N(N-1)} \{[N\pi_I](N\pi_J-1)+K_I[N\pi_J]\} & \text{if } K_I \le K_J , \\ \\ \frac{1}{N(N-1)} \{[N\pi_I](N\pi_J-1)+K_I[N\pi_J]+K_J-K_I\} & \text{if } K_I > K_J . \end{cases} \qquad (D.3)$$

In the case where π_I, π_J are integer multiples of N^{-1} we have $K_I = K_J = 0$ and so

$$\pi_{IJ} = \frac{\pi_I(N\pi_J-1)}{N-1} \quad \text{and} \quad \pi_{IJ} - \pi_I\pi_J = \frac{\pi_I(\pi_J-1)}{N-1} \quad \text{for } \pi_J \ge \pi_I \ .$$

The use of this last formula simplifies the calculation of the mean squared errors (4.2.34) and (4.2.36) and their estimators (4.2.35) and (4.2.37) very considerably. However it is necessary when using this approximation to ensure that the population is in size order with $\pi_J \ge \pi_I$ for all $J > I$.

APPENDIX E

CALCULATION OF P_{0C} FOR COLLOCATED SAMPLING

Without loss of generality we order the population such that $\pi_1 \le \pi_2 \le \ldots \le \pi_N$. Define K_I by $K_I = N\pi_I - [N\pi_I]$ and $D(I, \theta)$ by

$$D(I, \theta) = \begin{cases} [N\pi_I] & \text{if } K_I \le \theta , \\ [N\pi_I] + 1 & \text{if } K_I > \theta , \end{cases}$$

where θ is a random variable uniformly distributed on $[0, 1)$. Then the Ith population unit is in sample iff

$$L_I \le D(I, \theta) .$$

The requirement for an empty sample is therefore that $D(I, \theta) < L_I$ for all I. It can be shown that

$$D(1, \theta) \le D(2, \theta), \ldots, \le D(N, \theta) .$$

Hence

$$\{P_{0C}|\theta\} = \Pr\{L_I > D(I,\theta) \quad \text{for all} \quad I\}$$

$$\equiv \prod_{I=1}^{N} \{L_I > D(I,\theta) \mid L_H > D(H,\theta) \quad \text{for all} \quad H > I\}$$

$$= \frac{N-D(N,\theta)}{N} \cdot \frac{(N-1)-D\{(N-1),\theta\}}{N-1} \cdot \ldots$$

$$= \begin{cases} \frac{1}{N!} \prod_{I=1}^{N} \{I-D(I,\theta)\} & \text{if} \quad \min_I \{I, D(I,\theta)\} > 0 , \\ 0 & \text{otherwise.} \end{cases} \tag{E.1}$$

In the particular case where the $N\pi_I$ are all integers, $K_I = 0$ for all I and

$$P_{0C} = \begin{cases} \frac{1}{N!} \prod_{I=1}^{N} \left(I-N\pi_I\right) & \text{if} \quad \min_I \left(I-N\pi_I\right) > 0 , \\ 0 & \text{otherwise.} \end{cases} \tag{E.2}$$

In the general case it is necessary to evaluate

$$P_{0C} = \int_0^1 \{P_{0C}|\theta\} d\gamma ,$$

but since $D(I,\theta) \geq N\pi_I$ for all $\{I,\theta\}$ the formula (E.2) always gives an upper bound for P_{0C} .

BIBLIOGRAPHY

References marked * do not appear in text - see preface

*Agarwal, S.K., Kumar, P. and Dey, A. (1982). On unequal probability sampling of two units without replacement. *J. Roy. Statist. Soc. Ser. B* **44** (1), 43-46.

Aitchison, J. and Brown, J.A.C. (1957). *The Lognormal Distribution.* Cambridge University Press.

Ajgaonkar, S.G.P. (1965). On a class of linear estimators in sampling with varying probabilities without replacement. *J. Amer. Statist. Assoc.* **60**, 637-642.

Ajgaonkar, S.G.P. (1967). Unbiased estimators of the variance of the Narain, Horvitz and Thompson estimator. *Sankhyā Ser. B* **29**, 55-60.

Ajgaonkar, S.G.P. (1968). Bounds for the error variance of an estimator in sampling with varying probabilities from a finite population. *J. Amer. Statist. Assoc.* **63**, 91-98.

*Arnab, R. (1979a). An addendum to Singh and Singh's paper "On random non-response in unequal probability sampling". *Sankhyā Ser. C* **41** (2), 138-139.

*Arnab, R. (1979b). On strategies of sampling finite populations on successive occasions with varying probabilities. *Sankhyā Ser. C* **41** (2), 140-154.

*Arnab, R. (1981). Sampling on two occasions with varying probabilities. *Austral. J. Statist.* **23**, 360-364.

Asok, C. and Sukhatme, B.V (1976). On Sampford's procedure of unequal probability sampling without replacement. *J. Amer. Statist. Assoc.* **71**, 912-918.

Avadhani, M.S. and Srivastava, A.K. (1972). A comparison of Midzuno-Sen scheme with PPS sampling without replacement and its application to successive sampling. *Ann. Inst. Statist. Math.* **24**, 153-164.

Avadhani, M.S. and Sukhatme, B.V. (1967). Controlled sampling with varying probabilities with and without replacement. *Austral. J. Statist.* **9**, 8-15.

Basu, D. (1971). An essay on the logical foundations of survey sampling, Part one. *Foundations of Statistical Inference.* Holt, Rinehart and Winston. Edited by Godambe and Sprott.

Bayless, D.L. (1968). Variance estimation in sampling from finite populations. PhD Thesis, Texas A & M University, USA.

Bayless, D.L. and Rao, J.N.K. (1970). An empirical study of the stabilities of estimators and variance estimators in unequal probability sampling (n = 3 or 4). *J. Amer. Statist. Assoc.* **65**, 1645-1667.

Berg, S. (1974). A note on the RHC-method. *Scand. Actuarial J.* **57**, 108-114.

*Bhargava, N.K. (1978). On some applications of the technique of combined unordering. *Sankhyā Ser. C* **40** (1), 74-83.

*Bhargava, N.K., and Tikkiwal, B.D. (1978). On the axioms of sample formation with their relevance to *T*-classes of linear estimators. *Sankhyā Ser. C* **40** (1), 9-18.

*Biyani, Shiram H. (1980). On inadmissibility of the Yates-Grundy variance estimator in unequal probability sampling. *J. Amer. Statist. Assoc.* **75**, 709-712.

Bowley, A.L. (1913). Working class households in Reading. *J. Roy. Statist. Soc.* **76**, 672-701.

Brewer, K.R.W. (1963a). A model of systematic sampling with unequal probabilities. *Austral. J. Statist.* **5**, 5-13.

Brewer, K.R.W. (1963b). Ratio estimation and finite populations: some results deducible from the assumption of an underlying stochastic process. *Austral. J. Statist.* **5**, 93-105.

Brewer, K.R.W. (1967). A note on Fellegi's method of sampling without replacement with probabilities proportional to size. *J. Amer. Statist. Assoc.* **62**, 79-85.

Brewer, K.R.W. (1975). A simple procedure for sampling *πpswor* . *Austral. J. Statist.* **17** (3), 166-172.

Brewer, K.R.W. (1979). A class of robust sampling designs for large scale surveys. *J. Amer. Statist. Assoc.* **74** (4), 911-915.

Brewer, K.R.W., Early, L.J. and Joyce, S.F. (1972). Selecting several samples from a single population. *Austral. J. Statist.* **14**, 231-239.

Brewer, K.R.W., Early, L.J. and Hanif, M. (1980). Poisson, modified Poisson and collocated sampling. *J. Statist. Planning and Inference* (to appear).

Brewer, K.R.W. and Hanif, M. (1969a). Sampling without replacement with probability of inclusion proportional to size. I: Methods using Horvitz-Thompson estimator. Unpublished manuscript.

Brewer, K.R.W. and Hanif, M. (1969b). Sampling without replacement with probability of inclusion proportional to size. II: Methods using special estimators. Unpublished manuscript.

Brewer, K.R.W. and Hanif, M. (1970). Durbin's new multistage variance estimator. *J. Roy. Statist. Soc. Ser. B* **32**, 302-311.

Brewer, K.R.W. and Undy, G.C. (1962). Samples of two units drawn with unequal probabilities without replacement. *Austral. J. Statist.* **4**, 89-100.

Carroll, J.L. and Hartley, H.O. (1964). The symmetric method of unequal probability sampling without replacement. Abstract in *Biometrics* **20**, 908-909.

Cassel, Claes-M., Särndal, Carl-E., and Wretman, Jan H. (1976). Some results on generalized difference estimation and generalized regression estimation for finite populations. *Biometrika* **63**, 615-620.

Cassel, Claes-M., Särndal, Carl-E., and Wretman, Jan H. (1977). *Foundations of Inference in Survey Sampling*. John Wiley and Sons, Inc., New York.

Chaudhuri, A. (1971). Some sampling schemes to use Horvitz-Thompson estimator in estimating a finite population total. *Bull. Calcutta Statist. Assoc.* **20**, 37-66.

Chaudhuri, A. (1975). Some properties of estimators based on sampling schemes with varying probabilities. *Austral. J. Statist.* **17**, 22-28.

*Chaudhuri, A. (1976). On the choice of sampling strategies. *Calcutta Statist. Assoc. Bulletin* **25**, 119-128.

Chaudhuri, A. (1976). A simple method of sampling without replacement with inclusion probabilities exactly proportional to size. *Metrika* **22**, 147-152.

*Chaudhuri, A., and Adhikary, A.K. (1981). On optimality of double sampling strategies with varying probabilities. Preprint from *Applied Statistics*, Surveys and Computing Division, Indian Statistical Institute.

*Chaudhuri, A. and Arnab, R. (1978). On the role of sample size in determining efficiency of Horvitz-Thompson estimators. *Sankhyā Ser. C* **40** (2), 104-109.

*Chaudhuri, A. and Arnab, R. (1979). On the relative efficiencies of sampling strategies under a superpopulation model. *Sankhyā Ser. C* **41**, 40-43.

Chikkagoudar, M.S. (1966). A note on sampling with varying probabilities. *J. Indian Soc. Agric. Statist.* **18**, 86-92.

*Chikkagoudar, M.S. (1967). Sampling with preliminary random stratification. *Austral. J. Statist.* **9**, 57-70.

Chikkagoudar, M.S. (1969). Inverse sampling without replacement. *Austral. J. Statist.* **11**, 155-165.

Chotai, J. (1974). A note on the Rao-Hartley-Cochran method for PPS sampling over two occasions. *Sankhyā Ser. C* **36**, 173-180.

Choudhry, G.H. (1979). Selecting a sample of size n with *PPSWOR* from a finite population. *Survey Methodology* **5** (1), 79-95.

*Choudhry, G.H. and Singh, M.P. (1979). Sampling with unequal probabilities and without replacement - a rejective method. *Survey Methodology* **5** (2), 162-177.

Chromy, J.R. (1974). Pairwise probabilities in probability nonreplacement sampling. Presented at ASA meeting at St Louis, Missouri, USA.

Chromy, J.R. (1979). Sequential sample selection methods. Manuscript seen by courtesy of author.

Cochran, W.G. (1953, 1963, 1977). *Sampling Techniques*. John Wiley & Sons, Inc., New York.

Cochran, W.G. (1974). Two recent areas of sample survey research. In *A Survey of Statistical Design and Linear Models*. (J.N. Srivastava, ed.) North-Holland, Amsterdam, 101-115.

*Conner, J.R. and Welsch, D.E. (1968). Use of domain estimators with unequal probability in sample surveys. *J. Amer. Statist. Assoc.* **63**, 984-992.

Connor, W.S. (1966). An exact formula for the probability that two specified sample units will occur in a sample drawn with unequal probabilities and without replacement. *J. Amer. Statist. Assoc.* **61**, 384-390.

*Cumberland, W.G. and Royall, R.M. (1981). Prediction models and unequal probability sampling. *J. Roy. Statist. Soc. Ser. B* **43**, 353-367.

Das, A.C. (1951). On two phase sampling and sampling with varying probabilities. *Bull. Internat. Statist. Inst.* **XXXIII**, Book 2, 105-112.

Das, M.N. and Mohanty, S. (1973). On PPS sampling without replacement ensuring selection probabilities exactly proportional to sizes. *Austral. J. Statist.* **15**, 87-94.

*Das, A.K., and Tripathi, T.P. (1978). Use of auxiliary information in estimating the finite population variance. *Sankhyā Ser. C* **40** (2), 139-148.

Deming, W.E. (1960). *Sample Design in Business Research.* John Wiley & Sons, Inc., New York.

Dodds, D.J. and Fryer, J.G. (1971). Some families of selection probabilities for sampling proportional to size. *J. Roy. Statist. Soc. Ser. B* **33**, 263-274.

Durbin, J. (1953a). Some results in sampling when the units are selected with unequal probabilities. *J. Roy. Statist. Soc. Ser. B* **15**, 262-269.

Durbin, J. (1953b). Sampling with unequal probabilities of selection within strata. Unpublished to date; seen by courtesy of the author.

Durbin, J. (1967). Design of multistage surveys for the estimation of sampling errors. *Applied Statist.* **16**, 152-164.

Ecimovic, J.P. (1956). Three stage sampling with varying probabilities without replacement. *J. Indian Soc. Agric. Statist.* **8**, 14-44.

Fellegi, I.P. (1963). Sampling with varying probabilities without replacement: rotating and non-rotating samples. *J. Amer. Statist. Assoc.* **58**, 183-201.

Fellegi, I.P. and Winkworth, A.V. (1970). Survey methodology function in DBS. Technical Report, Dominion Bureau of Statistics, Canada.

Foreman, E.K. and Brewer, K.R.W. (1971). The efficient use of supplementary information in standard sampling procedures. *J. Roy. Statist. Soc. Ser. B* **33**, 391-400.

Fuller, W. (1971). A procedure for selecting nonreplacement unequal probabilities samples. Unpublished manuscript seen by courtesy of the author.

*Ghangurde, P.D. and Rao, J.N.K. (1969). Some results on sampling over two occasions. *Sankhyā Ser. A* **31**, 463-472.

Godambe, V.P. (1955). A unified theory of sampling from finite populations. *J. Roy. Statist. Soc. Ser. B* **17**, 269-278.

Godambe, V.P. and Joshi, V.M. (1965). Admissibility and Bayes estimation in sampling finite populations, I, II and III. *Ann. Math. Statist.* **36**, 1707-1742.

Godambe, V.P. and Thompson, M.E. (1977). Robust near optimal estimation in survey practice. *Bull. Internat. Statist. Inst.* **XLVII**, Book 3, 129-146.

Goodman, R. and Kish, L. (1950). Controlled selection - a technique in probability sampling. *J. Amer. Statist. Assoc.* **45**, 350-372.

Grundy, P.M. (1954). A method of sampling with probability exactly proportional to size. *J. Roy. Statist. Soc. Ser. B* **16**, 236-238.

Gupta, P.C. (1972). On some product type estimators in varying probability. *Austral. J. Statist.* **14**, 123-128.

Hájek, J. (1959). Optimum strategy and other problems in probability sampling. *Časopis. Pěst. Mat.* **84**, 387-423.

Hájek, J. (1958). Some contributions to the theory of probability sampling. *Bull. Internat. Statist. Inst.* **XXXVI**, Book 3, 127-134.

Hájek, J. (1964). Asymptotic theory of rejective sampling with varying probabilities from a finite population. *Ann. Math. Statist.* **35**, 1491-1523.

*Hájek, J. (1981). *Sampling from a Finite Population.* Marcel Dekker Inc., New York.

Hanif, M. (1969). Multistage cluster sampling. Unpublished MA Thesis, University of New South Wales.

Hanif, M. (1970a). Ratio estimation for multistage sampling. *Proceedings of Pak. Statist. Assoc.*, Vol. **15**, 2-6.

Hanif, M. (1970b). Comparison of variances estimation for unequal probabilities sampling. *J. Sci. Res. Punjab Univ.* **4**, 51-56.

Hanif, M. (1974). An asymptotic variance formula for sampling without replacement with unequal probabilities. *Libyan. J. Sci. Ser. A* **4**, 47-51.

Hanif, M. and Ahmad, M. (1977). Optimum level of clusterings for multipurpose sample surveys in Libya. *Bull. Internat. Statist. Inst.* **XLVII**, Book 4, 224-227.

*Hanif, M. and Brewer, K.R.W. (1977). Comparison of Poisson sampling with some selection procedures for unequal probabilities. *Punjab Univ. J. Math. (Lahore)* **10/11**, 33-39.

Hanif, M. and Brewer, K.R.W. (1979). Generalization of the Horvitz-Thompson estimator. *Punjab Univ. J. Math. (Lahore)* **12**

Hanif, M. and Brewer, K.R.W. (1980). On unequal probability sampling without replacement: a review. *Internat. Statist. Rev.* **48**, no. 3, 317-335.

Hansen, M.H. and Hurwitz, W.N. (1943). On the theory of sampling from a finite population. *Ann. Math. Statist.* **14**, 333-362.

Hansen, M.H. and Hurwitz, W.N. (1949). On the determination of optimum probabilities in sampling. *Ann. Math. Statist.* **20**, 426-432.

Hansen, M.H., Hurwitz, W.N. and Madow, W.G. (1953). *Sample Survey Methods and Theory.* John Wiley & Sons, Inc., New York.

Hanurav, T.V. (1962a). Some sampling schemes in probability sampling. *Sankhyā Ser. A* **24**, 421-428.

Hanurav, T.V. (1962b). On "Horvitz and Thompson estimator". *Sankhyā Ser. A* **24**, 429-436.

Hanurav, T.V. (1966). An optimum πps-sampling (Abstract). *Ann. Math. Statist.* **37**, 1859.

Hanurav, T.V. (1967). Optimum utilization of auxiliary information: πps sampling of two units from a stratum. *J. Roy. Statist. Soc. Ser. B* **29**, 374-391.

Hanurav, T.V. (1969). Optimum utilization of auxiliary information, πps sampling of two units from a stratum. (Addenda and corrigenda.) *J. Roy. Statist. Soc. Ser. B* **31**, 192-194.

Hartley, H.O. (1966). Systematic sampling with unequal probability and without replacement. *J. Amer. Statist. Assoc.* **61**, 739-748.

Hartley, H.O. and Chakrabarty, R.P. (1967). Evaluation of approximate variance formulas in sampling with unequal probabilities and without replacement. *Sankhyā Ser. B* **29**, 201-208.

Hartley, H.O. and Rao, J.N.K. (1962). Sampling with unequal probabilities and without replacement. *Ann. Math. Statist.* **33**, 350-374.

Hartley, H.O. and Rao, J.N.K. (1968). A new estimation theory for sample surveys. *Biometrika* **55**, 547-557.

Hartley, H.O., Rao, J.N.K. and Kiefer, G. (1969). Variance estimation with one unit per stratum. *J. Amer. Statist. Assoc.* **64**, 841-851.

Hege, V.S. (1965). An optimal property of the Horvitz and Thompson estimator. *J. Amer. Statist. Assoc.* **62**, 1013-1017.

*Ho, E.W.H. (1980). Model unbiasedness and the Horvitz-Thompson estimator in finite population sampling. *Austral. J. Statist.* **22**, 218-225.

Horvitz, D.G. and Thompson, D.J. (1952). A generalization of sampling without replacement from a finite universe. *J. Amer. Statist. Assoc.* **47**, 663-685.

*Isaki, C.T. and Fuller, W.A. (1982). Survey design under the regression superpopulation model. *J. Amer. Statist. Assoc.* **77**, 89-96.

Jessen, R.J. (1969). Some methods of probability non-replacement sampling. *J. Amer. Statist. Assoc.* **64**, 175-193.

Jessen, R.J. (1970). Probability sampling with marginal constraints. *J. Amer. Statist. Assoc.* **65**, 776-796.

*Joshi, V.M. (1969). Admissibility of estimates of the mean of a finite population. In *New Developments in Survey Sampling*. (N.L. Johnson and H. Smith, eds.) Wiley-Interscience, New York and London.

Kendall, M.G. and Stuart, A. (1966). *The Advanced Theory of Statistics*, Vol. 2. Griffin Publications.

Keyfitz, N. (1951). Sampling with probabilities proportional to size: adjustment for changes in the probabilities. *J. Amer. Statist. Assoc.* **46**, 105-109.

Kish, L. (1965). *Survey Sampling*. J. Wiley & Sons, Inc., New York.

Konijn, H.S. (1973). On modified schemes of systematic sampling with several random starts. *Commun. Statist.* **1**, 133-148.

*Koop, J.C. (1963). On the axioms of sample formation and their bearing on the construction of linear estimators in sampling theory for finite populations. *Metrika* **7** (2 and 3), 81-114, 165-204.

*Koop, J.C. (1978). Comments on the paper by Bhargava and Tikkiwal. *Sankhyā Ser. C* **40** (1), 19-20.

Lahiri, D.B. (1951). A method of sample selection providing unbiased ratio estimates. *Bull. Internat. Statist. Inst.* **XXXIII**, Book 2, 133-140.

Lanke, J. (1974). On non-negative variance estimators in survey sampling. *Sankhyā Ser. C* **36** (1), 33-42.

Liu, T.P. (1973). A short note on the unbiasedness of Sen-Yates-Grundy variance estimator. *Tamkang J. Math.* **4**, 45-47.

Madow, W.G. (1949). On the theory of systematic sampling, II. *Ann. Math. Statist.* **20**, 333-354.

Midzuno, H. (1950). An outline of the theory of sampling systems. *Ann. Inst. Statist. Math.* **1**, 149-156.

Midzuno, H. (1952). On the sampling system with probability proportionate to sum of sizes. *Ann. Inst. Statist. Math.* **3**, 99-107.

Mukhopadhyay, P. (1972). A sampling scheme to realise a pre-assigned set of inclusion probabilities of first two orders. *Bull. Calcutta Statist. Assoc.* **21**, 87-122.

*Mukhopadhyay, P. (1977). Robust estimators of finite population total under certain linear regression models. *Sankhyā Ser. C* **39** (1), 71-87.

Murthy, M.N. (1957). Ordered and unordered estimators in sampling without replacement. *Sankhyā* **18**, 379-390.

*Murthy, M.N. (1967). *Sampling Theory and Methods.* Statist. Pub. Soc. Calcutta.

Narain, R.D. (1951). On sampling without replacement with varying probabilities. *J. Indian Soc. Agric. Statist.* **3**, 169-175.

Neyman, J. (1934). On the two different aspects of the representative method: the method of stratified sampling and the method of purposive selection. *J. Roy. Statist. Soc.* **97**, 558-606.

Nisselson, Harold and Isaki, Cary T. (1977). Statistical theory and practice at the US Bureau of the Census. *Bull. Internat. Statist. Inst.* **XLVII**, Book 3, 147-158.

Ogus, J.L. (1969). A note on the "Necessary Best Estimator". *J. Amer. Statist. Assoc.* **64**, 1350-1352.

Ogus, J.L. and Clark, D.F. (1971). The annual survey of manufacturers: a report on methodology. Technical Paper No. 24, US Bureau of the Census, Washington, D.C.

*Padmawar, V.R. (1981). A note on the comparison of certain sampling strategies. *J. Roy. Statist. Soc. Ser. B* **43**, 321-326.

*Patel, H.C. and Dharmadhikari, S.W. (1978). Admissibility of Murthy's and Midzuno's estimators within the class of linear unbiased estimators of finite population totals. *Sankhyā Ser. C* **40** (1), 21-28.

Pathak, P.K. (1961). Use of "Order Statistics" in sampling without replacement. *Sankhyā Ser. A* **23**, 409-414.

Pathak, P.K. (1962). On sampling with unequal probabilities. *Sankhyā Ser. A* **24**, 315-326.

Pathak, P.K. (1964). On inverse sampling with unequal probabilities. *Biometrika* **51**, 185-193.

Pathak, P.K. (1966). An estimator in PPS sampling for multiple characteristics. *Sankhyā Ser. A* **28**, 35-40.

Pathak, P.K. (1967a). Asymptotic efficiency of Des Raj's strategy - I. *Sankhyā Ser. A* **29**, 283-298.

Pathak, P.K. (1967b). Asymptotic efficiency of Des Raj's strategy - II. *Sankhyā Ser. A* **29**, 299-304.

Pathak, P.K. and Shukla, N.D. (1966). Non-negativity of a variance estimator. *Sankhyā Ser. A* **28**, 41-46.

Patterson, H.D. (1950). Sampling on successive occasions with partial replacement of units. *J. Roy. Statist. Soc. Ser. B* **12**, 241-255.

Pedgaonkar, A.M. and Ajgaonkar, S.G.P. (1978). Comparison of sampling strategies. *Metrika* **25**, 149-153.

Raj, D. (1954a). Ratio estimation in sampling with equal and unequal probabilities. *J. Indian Soc. Agric. Statist.* **6**, 127-138.

Raj, D. (1954b). On sampling with probabilities proportional to size. *Ganita* **5**, 175-182.

Raj, D. (1956a). Some estimators in sampling with varying probabilities without replacement. *J. Amer. Statist. Assoc.* **51**, 269-284.

Raj, D. (1956b). A note on the determination of optimum probabilities in sampling without replacement. *Sankhyā* **17**, 197-200.

Raj, D. (1958). On the relative accuracy of some sampling techniques. *J. Amer. Statist. Assoc.* **53**, 98-101.

Raj, D. (1964). The use of systematic sampling with probability proportionate to size in a large scale survey. *J. Amer. Statist. Assoc.* **59**, 251-255.

Raj, D. (1965). Variance estimation in randomized systematic sampling with probability proportionate to size. *J. Amer. Statist. Assoc.* **60**, 278-284.

Raj, D. (1966). Some remarks on a simple procedure of sampling without replacement. *J. Amer. Statist. Assoc.* **61**, 391-396.

Ramachandran, G. and Rao, T.J. (1974). Allocation to strata and relative efficiencies of stratified and unstratified πps samples. *J. Roy. Statist. Soc. Ser. B* **36**, 292-298.

Ramakrishnan, M.K. (1973). An alternative proof of the admissibility of the Horvitz-Thompson estimator. *Ann. Statist.* **1**, 577-579.

Ramakrishnan, M.K. (1975). A generalisation of the Yates-Grundy variance estimator. *Sankhyā Ser. C* **37**, 204-206.

*Rao, C.R. (1977). Some problems of sample surveys. *Sankhyā Ser. C* **39** (3), 128-139.

*Rao, G.D. (1978). Double PPS sample to make design self-weighting with equal number of SSU's. *Sankhyā Ser. C* **40** (2), 177-189.

Rao, J.N.K. (1961a). On the estimate of the variance in unequal probability sampling. *Ann. Inst. Statist. Math.* **13** (1), 57-60.

Rao, J.N.K. (1961b). On sampling with varying probabilities and with replacement in subsampling designs. *J. Indian. Soc. Agric. Statist.* **13**, 211-217.

Rao, J.N.K. (1962). On the estimation of relative efficiency of sampling procedures. *Ann. Inst. Statist. Math.* **14**, 143-150.

Rao, J.N.K. (1963a). On two systems of unequal probability sampling without replacement. *Ann. Inst. Statist. Math.* **15**, 67-72.

Rao, J.N.K. (1963b). On three procedures of unequal probability sampling without replacement. *J. Amer. Statist. Assoc.* **58**, 202-215.

Rao, J.N.K. (1965). On two simple schemes of unequal probability sampling without replacement. *J. Indian Statist. Assoc.* **3**, 173-180.

Rao, J.N.K. (1966a). Alternative estimators in pps sampling for multiple characteristics. *Sankhyā Ser. A* **28**, 47-60.

Rao, J.N.K. (1966b). On the relative efficiency of some estimators in pps sampling for multiple characteristics. *Sankhyā Ser. A* **28**, 61-70.

Rao, J.N.K. (1966c). On the comparison of sampling with and without replacement. *Rev. Internat. Statist. Inst.* **34**, 125-138.

Rao, J.N.K. (1975). Unbiased variance estimation for multistage sampling designs. *Sankhyā Ser. C* **37**, 133-139.

Rao, J.N.K. (1978). Sampling designs involving unequal probabilities of selection and robust estimation of a finite population total. *Contributions to Survey Sampling and Applied Statistics. Papers in honor of H.O. Hartley.* Edited by H.A. David. Academic Press, 69-87.

Rao, J.N.K. (1979). On deriving mean square errors and their non-negative unbiased estimators in finite population sampling. *J. Indian Statist. Assoc.* **17**, 125-136. (Huzurbazar felicitation volume.)

Rao, J.N.K. and Bayless, D.L. (1969). An empirical study of the stabilities of estimators and variance estimators in unequal probability sampling of two units per stratum. *J. Amer. Statist. Assoc.* **64**, 540-549.

Rao, J.N.K., Hartley, H.O. and Cochran, W.G. (1962). On a simple procedure of unequal probability sampling without replacement. *J. Roy. Statist. Soc. Ser. B* **24**, 482-491.

Rao, J.N.K. and Singh, M.P. (1973). On the choice of estimator in survey sampling. *Austral. J. Statist.* **15**, 95-104.

Rao, J.N.K. and Vijayan, K. (1977). On estimating the variance in sampling with probability proportional to aggregate size. *J. Amer. Statist. Assoc.* **72**, 579-584.

*Rao, T.J. (1967). On the choice of strategy for ratio method of estimation. *J. Roy. Statist. Soc. Ser. B* **29**, 392-397.

Rao, T.J. (1970). Horvitz and Thompson and Des Raj estimators revisited. Research Report, University of Manchester.

Rao, T.J. (1971). πps sampling designs and the Horvitz-Thompson estimator. *J. Amer. Statist. Assoc.* **66**, 872-875.

*Rao, T.J. (1977). Optimum allocation of sample size and prior distributions: a review. *Internat. Statist. Rev.* **45**, 173-180.

Reddy, V.N. and Rao, T.J. (1977). Modified PPS method of estimation. *Sankhyā Ser. C* **39** (3), 185-197.

*Rosén, B. (1972). Asymptotic theory for successive sampling with varying probabilities without replacement I and II. *Ann. Math. Statist.* **43**, 373-397 and 748-776.

*Rosén, B. (1974). Asymptotic theory for Des Raj's estimator, I and II. *Scand. J. Statist.* **1**, 71-83 and 135-144.

Roy, J. (1957). A note on estimation of variance components in multistage sampling with varying probabilities. *Sankhyā* **17**, 367-372.

Roy, J. and Chakravarti, I.M. (1960). Estimating the mean of a finite population. *Ann. Math. Statist.* **31**, 392-398.

*Royall, R.M. (1970). On finite population sampling theory under certain linear regression models. *Biometrika* **57**, 377-387.

Royall, R.M. and Herson, J. (1973a). Robust estimation in finite populations I. *J. Amer. Statist. Assoc.* **68**, 880-889.

Royall, R.M. and Herson, J. (1973b). Robust estimation in finite populations II. Stratification on a size variable. *J. Amer. Statist. Assoc.* **68**, 890-893.

Sadasivan, G. and Sharma, S. (1974). Two dimensional varying probability sampling without replacement. *Sankhyā Ser. C* **36** (3), 157-166.

Samiuddin, M. (1981). A simple procedure of unequal probability sampling. *Biometrika* **68** (3), 728-731.

Samiuddin, M., Hanif, M. and Asad, H. (1978). Some remarks on PPS sampling without replacement. Presented at the Fourth Australian Statistical Conference, Canberra.

Sampford, M.R. (1962). Methods of cluster sampling with and without replacement for clusters of unequal sizes. *Biometrika* **49**, 27-40.

Sampford, M.R. (1967). On sampling without replacement with unequal probabilities of selection. *Biometrika* **54**, 499-513.

*Sampford, M.R. (1969). A comparison of some possible methods of sampling from smallish populations, with units of unequal size. In *New Developments in Survey Sampling* (N.L. Johnson and H. Smith, eds.) Wiley-Interscience, New York and London.

Sampford, M.R. (1975). The Horvitz and Thompson in theory and practice: A historical paper. Unpublished.

*Sanchez-Crespo, J.L. (1977). A new sampling scheme: selection with graduated variable probabilities without replacement. *Bull. Internat. Statist. Inst.* **XLVII**, Book 4, 458-461.

Sankaranarayanan, K. (1969). An IPPS sampling scheme using Lahiri's method of selection. *J. Indian Soc. Agric. Statist.* **21**, 58-66.

*Schreuder, H.T. (1974). Two stratified sampling alternatives to 3-P sampling. Proceedings: *Monitoring Forest Environment Through Successive Sampling*. (T. Cunia, ed.)

*Schreuder, H.T. (1982). Poisson sampling and some alternatives in forestry. Submitted to *J. Amer. Statist. Assoc.*

*Schreuder, H.T., Sedransk, J. and Ware, K.D. (1968). 3-P sampling and some alternatives, I. *Forest Science* **14**, 429-454.

*Schreuder, H.T., Sedransk, J. and Ware, K.D. (1971). Forest sampling in the absence of a well-defined frame. In *Statistical Ecology*, Vol. 2: *Sampling and Modeling Biological Populations and Population Dynamics*, 119-150. Pennsylvania State University Press.

*Schreuder, H.T., Sedransk, J., Ware, K.D. and Hamilton, D.A. (1971). 3-P sampling and some alternatives. *Forest Science* **17**, 103-118.

Scott, A.J., Brewer, K.R.W. and Ho, E.W.H. (1978). Finite population sampling and robust estimation. *J. Amer. Statist. Assoc.* **73**, 359-361.

Sen, A.R. (1953). On the estimate of the variance in sampling with varying probabilities. *J. Indian Soc. Agric. Statist.* **5**, 119-127.

Sen, A.R. (1955). A simple design in sampling with varying probabilities. *J. Indian Soc. Agric. Statist.* **7**, 57-69.

Sen, A.R. (1976). Developments in PPS sampling, impact on current research. Presented at the International Biometric Conference, Boston.

Seth, G.R. (1966). On estimators of variance of estimate of population total in varying probabilities. *J. Indian Soc. Agric. Statist.* **18** (2), 52-56.

Sethi, V.K. (1962). Some consequences of an interpretation of varying probability sampling. *Sankhyā Ser. B* **24**, 215-222.

Sethumadhavi, R. and Pajagopalan, M. (1974). Use of PPS with 3-P sampling procedure. *Sankhyā Ser. C* **36** (2), 167-172.

Singh, D. (1954). On the efficiency of the sampling with varying probabilities without replacement. *J. Indian. Soc. Agric. Statist.* **6**, 48-57.

*Singh, M.P. (1978). Alternative estimators in pps sampling. *Survey Methodology* **4**, 264-280.

Singh, P. (1978). A sampling scheme with inclusion probability proportional to size. *Sankhyā Ser. C* **40**, 122-128.

*Singh, R. (1972). On Pathak and Rao's estimate in pps with replacement over two occasions. *Sankhyā Ser. A* **34**, 301-303.

Singh, R. (1975a). A note on optimum stratification in sampling with varying probabilities. *Austral. J. Statist.* 17, 12-21.

Singh, R. (1975b). A note on the efficiency of ratio estimate with Midzuno's scheme of sampling. *Sankhyā Ser. C* 37, 211-214.

Singh, R. and Lal, M. (1978). On the construction of random groups in the RHC scheme. *Sankhyā Ser. C* 40, 129-135.

Singh, R. and Kishore, L. (1975). On Rao, Hartley and Cochran's method of sampling. *Sankhyā Ser. C* 37, 88-94.

Singh, R. and Singh, B. (1974). On replicated samples drawn with Rao, Hartley and Cochran's scheme. *Sankhyā Ser. C* 36, 147-150.

*Singh, R. and Sukhatme, B.V. (1972). Optimum stratification in sampling with varying probabilities. *Ann. Inst. Statist. Math.* 24, 485-494.

*Singh, S. and Singh, R. (1979). On random non-response in unequal probability sampling. *Sankhyā Ser. C* 41, 127-137.

Sinha, B.K. (1973). On sampling schemes to realize preassigned sets of inclusion probabilities of first two orders. *Bull. Calcutta Statist. Assoc.* 22, 89-110.

Smith, H.F. (1938). An experimental law describing heterogeneity in the yields of agricultural crops. *J. Agric. Science* 28, 1-23.

*Solomon, H. and Zacks, S. (1970). Optimal design of sampling from finite populations. A critical review and indication of new research areas. *J. Amer. Statist. Assoc.* 65, 653-677.

*Srinath, K.P. and Hidiroglou, M.A. (1977). Estimation of variance in multi-stage sampling. *Survey Methodology* 3, 104-113.

Stevens, W.L. (1958). Sampling without replacement with probability proportional to size. *J. Roy. Statist. Soc. Ser. B* 20, 393-397.

Stuart, A. (1964). Multistage sampling with preliminary random stratification of first stage units. *Rev. Internat. Statist. Inst.* 32, 193-201.

Sukhatme, P.V. and Sukhatme, B.V. (1970). *Sampling Theory of Surveys with Applications*. Asian Publishing House, Calcutta.

Sunter, A.B. (1977a). Response burden, sample rotation, and classification renewal in economic surveys. *Internat. Statist. Rev.* 45, 209-222.

*Sunter, A.B. (1977b). List sequential sampling with equal or unequal probabilities without replacement. *Applied Statistics* 26, 261-268.

*Tikkiwal, B.D. (1966). On T_2 and T_1 classes of estimators in sampling with varying probabilities and without replacement. *J. Indian Soc. Agric. Statist.* 18 (1), 62-72.

*Tripathi, T.P. and Srivastava, O.P. (1979). Estimation on successive occasions using PPS WR sampling. *Sankhyā Ser. C* 41, 84-91.

Vijayan, K. (1966). On Horvitz-Thompson and Des Raj estimators. *Sankhyā Ser. A* 28, 87-92.

Vijayan, K. (1968). An exact πps sampling scheme; generalization of a method of Hanurav. *J. Roy. Statist. Soc. Ser. B* **30**, 556-566.

Vijayan, K. (1975). On estimating the variance in unequal probability sampling. *J. Amer. Statist. Assoc.* **70**, 713-716.

Vos, J.W.E. (1974). Sampling with unequal probabilities. *Statistica Neerlandica* **28**, 11-49 and 69-107.

Yates, F. (1949, 1953, 1960, 1981). *Sampling for Censuses and Surveys*. Griffin, London.

Yates, F. and Grundy, P.M. (1953). Selection without replacement from within strata with probability proportional to size. *J. Roy. Statist. Soc. Ser. B* **15**, 253-261.

Yamamoto, S. (1955). On the theory of sampling with probabilities proportionate to given values. *Ann. Inst. Statist. Math.* **7**, 25-38.

Zarkovic, S.S. (1960). On the efficiency of sampling with various probabilities and the selection of the units with replacement. *Metrika* **3**, 53-60.

SELECTION PROCEDURES' INDEX

AUTHORS' INDEX

GPSR Compliance

The European Union's (EU) General Product Safety Regulation (GPSR) is a set of rules that requires consumer products to be safe and our obligations to ensure this.

If you have any concerns about our products, you can contact us on ProductSafety@springernature.com

In case Publisher is established outside the EU, the EU authorized representative is:

Springer Nature Customer Service Center GmbH
Europaplatz 3
69115 Heidelberg, Germany

Batch number: 10704701

Printed by Printforce, the Netherlands